Applied Complex Variables

HARPERCOLLINS COLLEGE OUTLINE

Applied Complex Variables

Paul C. DuChateau, Ph.D.
Colorado State University

 For information address HarperCollins Publishers, Inc., 10 East 53rd Street, New York, NY, 10022.

An American BookWorks Corporation Production

Project Manager: William Hamill
Editor: Swayse Editorial

Library of Congress Cataloging-in-Publication Data

DuChateau, Paul C.
Applied complex variables / Paul C. DuChateau.
p. cm. —(HarperCollins college outline series)
Includes index.
ISBN: 0-06-467152-6 (pbk.)
1. Functions of complex variables. I. Title. II. Series.
QA331.7.D83 1993
515'.9—dc20 92-54432

93 94 95 96 97 ABW/RRD 10 9 8 7 6 5 4 3 2 1

Contents

Preface

The theory of functions of a complex variable has long been a fundamental part of mathematical analysis. It is a subject that displays an elegant formal structure while its methods provide powerful tools for attacking a wide range of applied problems. This book surveys the subject at an introductory level and is intended to supplement any of the manor undergraduate level texts in complex variables. It may also be used for a program of self study or even as a primary source in an undergrduate introduction to functions of a complex variable.

This book has been written with two objectives. One of the goals is to present the theory of functions of a complex variable to an audience having only the standard calculus background. Chapters 1 through 4 contain this presentation, arranged as follows:

Chapter 1 The algebra of complex numbers

Chapter 2 Derivatives
Chapter 3 Integrals
Chapter 4 Power Series
} Functions of a Complex Variable

Each chapter contains key theorems and their proofs together with numerous examples selected to illustrate the concepts. Supplementary problems with answers are provided so the students may test their grasp of the material. The treatment is complete but presented at a level aimed at students in science and engineering whose mathematical background does not extend beyond the usual courses in calculus.

To present the many varied applications of complex variable theory is a second objective of this book. This material should be of interest not only to applied scientists but also to mathematicians curious about the practical uses of this elegant subject. Chapter 5 (The Residue Theorem and the Argument Principle) and chapter 6 (Conformal Mapping) provide additional foundation for the applications which are developed in chapters 7 and 8.

Chapter 7 shows how the residue theorem may be used in the inversion of integral transforms which are, in turn, used to solve a variety of problems in ordinary and partial differential equations. In addition the argument principle is used to locate the zeroes of complex polynomials for the purpose of detecting instability in linear systems.

Chapter 8 contains a treatment of two dimensional potential theory and begins with a discussion of vector fields V which are simultaneously solenoidal (div V = 0) and irrotational (cur V = 0). Such fields occur naturally in the description of various steady state physical phenomena including electrostatics, magnetostatics, and steady incompressible irrotational fluid flow. When the vector fields are 2-dimensional the techniques of complex function theory can be brought to bear to yield useful and often surprising results. Problem 8.9, for example, provides an explanation of what causes the curve in a curve ball. The conformal mapping concepts of chapter 6 are also used extensively in this discussion of potential theory.

I would like to thank my wife Sara and daughter Danielle for bearing up during the rather intense period when this book was written. I would also like to express my appreciation to Fred N. Grayson for giving me the opportunity to be a part of this project, and to Bill Hamill for guiding the book through its various stages of development.

Paul C. DuChateau

1

Complex Numbers

The development of complex numbers is motivated by the fact that there are polynomial equations having real coefficients but no real solutions. However, it is possible to enlarge the real number system in such a way that all such equations are solvable within the enlarged set of numbers. This new system, called the complex number system, contains the real numbers as a subset and the complex definitions for addition and multiplication reduce to the familiar real operations when applied to complex numbers that are real.

In this chapter we examine the arithmetic of complex numbers. After first introducing complex numbers as ordered pairs we adopt two alternative notations for complex numbers, namely the Cartesian and polar representations. These modes of representation are convenient for computational purposes. The ordered pair notation suggests identifying complex numbers with points in the plane and we illustrate ways in which this identification can be used to advantage.

COMPLEX NUMBERS

It is easy to see that there are no real numbers x that satisfy $x^2 + 1 = 0$. The fact that there are polynomial equations having real coefficients but no real solutions could be viewed as an inadequacy in the real numbers. In order to remedy this situation we enlarge the number system.

DEFINITION

We define a *complex number* z to be an ordered pair $z = (a, b)$ of real numbers. Every real number r can be identified with the unique complex number $(r, 0)$. Thus we can think of the real numbers as a subset of the complex numbers.

REAL AND IMAGINARY PART

Complex numbers of the form $(0, q)$ are referred to as *imaginary numbers* and we refer to p and q as the *real and imaginary parts* of the complex number $z = (p, q)$. Sometimes we use the notation $\operatorname{Re} z = p$ and $\operatorname{Im} z = q$ for the real and imaginary parts of z. We say that two complex numbers are equal and write $z_1 = z_2$, if and only if $\operatorname{Re} z_1 = \operatorname{Re} z_2$ and $\operatorname{Im} z_1 = \operatorname{Im} z_2$.

COMPLEX ARITHMETIC

We define the *sum* and *product* of complex numbers $z_1 = (a_1, b_1)$, $z_2 = (a_2, b_2)$ as follows:

$$z_1 + z_2 = (a_1 + a_2, b_1 + b_2) \tag{1.1}$$

$$z_1 z_2 = (a_1 a_2 - b_1 b_2, a_1 b_2 + a_2 b_1) \tag{1.2}$$

Note that when these operations are restricted to the real numbers, they reduce to addition and multiplication in the usual (real) sense; i.e.,

$$(a_1, 0) + (a_2, 0) = (a_1 + a_2, 0) \quad \text{and} \quad (a_1, 0)\,(a_2, 0) = (a_1 a_2, 0)\,.$$

Thus the complex numbers are an extension of the real numbers. Note also that for any real number r, and complex number $z = (a, b)$ we have $rz = (r, 0)\,(a, b) = (ra, rb)$.

THE IMAGINARY UNIT

The imaginary number $(0, 1)$, referred to as the *imaginary unit*, satisfies $(0, 1) \cdot (0, 1) = (-1, 0)$; i.e., $(0, 1)$ is the square root of -1. We denote this number by the symbol i (some texts use j in place of i). Then we can introduce an alternative notation for complex numbers: $z = (a, b) = a + ib$. This notation has obvious advantages for doing complex arithmetic.

$$\begin{aligned} z_1 + z_2 = (a_1, b_1) + (a_2, b_2) &= a_1 + ib_1 \\ &\underline{\;\; a_2 + ib_2 \;\;} \\ a_1 + a_2 + i\,(b_1 + b_2) &= (a_1 + a_2, b_1 + b_2) \end{aligned}$$

$$\begin{aligned} z_1 z_2 = (a_1, b_1)\,(a_2, b_2) &= a_1 + ib_1 \\ &\underline{\times\, a_2 + ib_2} \\ &a_1 a_2 + i a_2 b_1 \\ &\underline{\quad i a_1 b_2 + i^2 b_1 b_2} \qquad (i^2 = -1) \\ &(a_1 a_2 - b_1 b_2) + i\,(a_1 b_2 + a_2 b_1) \end{aligned}$$

COMPLEX CONJUGATE

For complex number $z = a + ib$ we define the *complex conjugate* of z to be the complex number $z^* = a - ib$. Then the conjugate has the following properties

$$(z_1 + z_2)^* = z_1^* + z_2^*, \quad (z_1 z_2)^* = z_1^* z_2^*, \text{ and } (z^*)^* = z \tag{1.3}$$

Also

$$z + z^* = 2\operatorname{Re} z, \quad z - z^* = 2\operatorname{Im} z \text{ and } zz^* = a^2 + b^2 = |z|^2 \tag{1.4}$$

The real number $|z| = (zz^*)^{1/2}$ is called the *modulus* of the complex number z. The only complex number with modulus equal to zero is the number (0, 0).

DIVISION WITH COMPLEX NUMBERS

For complex numbers z_1, z_2 with $|z_2|$ not equal to 0 we have

$$\frac{z_1}{z_2} = \frac{z_1 z_2^*}{z_2 z_2^*} = \frac{1}{|z_2^*|^2} z_1 z_2^* = \left(\frac{a_1 a_2 + b_1 b_2}{|z_2|^2}, \frac{a_2 b_1 - a_1 b_2}{|z_2|^2} \right)$$

Division by $z = (0, 0)$ is not defined just as division by $x = 0$ is not defined for real numbers.

ALGEBRAIC PROPERTIES OF THE COMPLEX NUMBERS

Complex addition and multiplication are commutative, associative and distributive like their counterparts for the real numbers. The real number 0 is the additive identity in the reals and is identified with the complex number (0, 0) which is the additive identity in the complex numbers. Also the real number 1, the multiplicative identity for the reals, is identified with the complex number (1, 0) which is the multiplicative identity for the complex numbers. Every complex number $z = (a, b)$ has an additive inverse $-z = (-a, -b)$ and, (for $|z| > 0$), a multiplicative inverse $z^{-1} = z^*/|z|^2$. The complex numbers form a field but not an ordered field. That is, $z_1 < z_2$ has no meaning when z_1 and z_2 are not real numbers.

Complex Numbers as Vectors

We can identify complex numbers with 2-vectors in the plane as shown in Figure 1.1. We refer to the plane as the *complex plane* and speak of the x and y axes as the *real* and *imaginary axes*, respectively. Then $|z|$, the modulus of the complex number z, can be interpreted as the length of the vector z. In this context the following inequality is geometrically evident

$$|z_1 + z_2| \le |z_1| + |z_2| \quad \text{for all complex numbers } z_1, z_2 \tag{1.5}$$

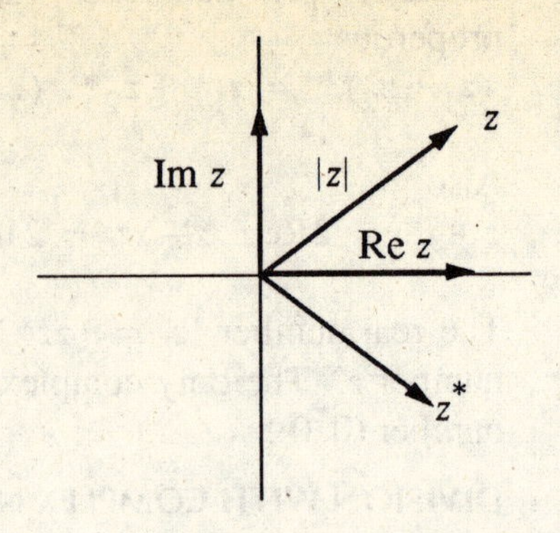

Figure 1.1

(1.5) is known as the *triangle inequality.* Note that $|z_1 + z_2|$ is equal to $|z_1| + |z_2|$ if and only if $z_1 + z_2$, z_1 and z_2 are colinear.

POLAR REPRESENTATION

The vector interpretation for complex numbers suggest the following alternative way of representing complex numbers. For nonzero complex number $z = a + ib$ write

$$z = r(\cos\vartheta + i\sin\vartheta) \quad \text{where } r = |z| \text{ and } \tan\vartheta = b/a \tag{1.6}$$

We refer to this as the *polar form* of the complex number z. Here ϑ is called the *argument* of z and we write $\vartheta = \operatorname{Arg} z$. It is customary to take the principle value for ϑ; i.e., $-\pi < \vartheta \le \pi$. Note that the polar form of the complex conjugate is then $z^* = r(\cos\vartheta - i\sin\vartheta)$.

Theorem 1.1 For nonzero complex numbers z_1 and z_2

$$z_1 z_2 = r_1 r_2 (\cos(\vartheta_1 + \vartheta_2) + i\sin(\vartheta_1 + \vartheta_2)) \tag{1.7}$$

and,

$$z_1/z_2 = r_1/r_2 (\cos(\vartheta_1 - \vartheta_2) + i\sin(\vartheta_1 - \vartheta_2)) \tag{1.8}$$

For any nonzero complex number z, and for any positive integer n

$$z^n = r^n(\cos n\vartheta + i\sin n\vartheta) \tag{1.9}$$

and

$$z^{1/n} = r^{1/n}(\cos((\vartheta + 2k\pi)/n) + i\sin((\vartheta + 2k\pi)/n)) \tag{1.10}$$

for $k = 0$ to $n - 1$.

EULER'S FORMULA AND DEMOIVRE'S THEOREM

We define

$$e^{i\vartheta} = \cos\vartheta + i\sin\vartheta \tag{1.11}$$

Then $e^{i\vartheta_1}e^{i\vartheta_2} = e^{i(\vartheta_1+\vartheta_2)}$ follows from (1.7) in the case $r_1 = r_2 = 1$ and thus (1.11) is formally consistent with the usual rule of real exponents. It follows from (1.9) that for positive integers n,

$$e^{in\vartheta} = (\cos\vartheta + i\sin\vartheta)^n = \cos n\vartheta + i\sin n\vartheta \tag{1.12}$$

The results (1.11) and (1.12) are known respectively as *Euler's formula* and *deMoivre's theorem.* Note that Euler's formula simplifies the polar representation (1.6) to $z = re^{i\vartheta}$. Then

$$z_1z_2 = r_1r_2e^{i(\vartheta_1+\vartheta_2)} \quad \text{and} \quad z_1/z_2 = (r_1/r_2)\,e^{i(\vartheta_1-\vartheta_2)}$$

$$z^n = r^ne^{in\vartheta} \quad \text{and} \quad z^{1/n} = r^{1/n}e^{i(\vartheta+2\pi k)/n} \quad \text{for } k = 0 \text{ to } n-1.$$

Sets in the Complex Plane

NEIGHBORHOODS

Let z_0 denote a fixed complex number. Then for real number $\varepsilon > 0$, the set of points in the complex plane that satisfy $|z - z_0| < \varepsilon$ is a disc of radius ε with center at z_0. We refer to this set as an ε *– neighborhood* (or neighborhood) of z_0 and denote the neighborhood by $N_\varepsilon(z_0)$.

INTERIOR POINTS, OPEN SETS

Let D denote a set of complex numbers. If z belongs to D and there exists a neighborhood of z containing only points of D then we say z is an *interior point* of D. If all points of D are interior points then we say D is an *open set.*

BOUNDARY POINTS, CLOSED SETS

A set D in the complex plane is *bounded* if the set is contained within a disc of sufficiently large radius. A point z (not necessarily in D) such that every neighborhood of z contains points of D and points not in D is said to be a *boundary point* of D. If the set D contains all of its boundary points then we say D is a *closed set.* Note that closed and open are not opposites since there are sets which are both open and closed and there are other sets which are neither open nor closed.

DOMAINS

The set D is *connected* if any two points in D can be joined by a polygonal path consisting of finitely many segments each of which contains only points of D. An open connected set is called a *domain* or a *region* in the complex plane.

SOLVED PROBLEMS

PROBLEM 1.1

For complex numbers $z_1 = (2, 1)$, $z_2 = (-1, 5)$ and $z_3 = (4, -3)$ compute:

(a) $z_1 + z_2$

(b) $z_1 z_3$

(c) $z_2 z_3$

(d) $(z_1 + z_2) z_3$

SOLUTION 1.1

We begin by writing $z_1 = 2 + i$, $z_2 = -1 + 5i$ and $z_3 = 4 - 3i$. Then

(a)
$$\begin{array}{r} 2 + i \\ \underline{-1 + 5i} \\ 1 + 6i \end{array}$$

(b)
$$\begin{array}{r} 2 + i \\ \underline{\times 4 - 3i} \\ 8 + 4i \\ \underline{-6i - 3i^2} \\ 11 - 2i \end{array}$$

(c)
$$\begin{array}{r} -1 + 5i \\ \underline{\times 4 - 3i} \\ -4 + 20i \\ \underline{3i - 15i^2} \\ 11 + 23i \end{array}$$

(d)
$$\begin{array}{r} 1 + 6i \\ \underline{\times 4 - 3i} \\ 4 + 24i \\ \underline{-3i - 18i^2} \\ 22 + 21i \end{array}$$

Note that the result (d) equals the sum of the results from (b) and (c); i.e.,

$$z_1 z_3 + z_2 z_3 = (z_1 + z_2) z_3$$

This is a demonstration of the distributive property of complex multiplication and addition.

PROBLEM 1.2

For complex numbers $z_1 = (2, 1)$, $z_2 = (-1, 5)$ and $z_3 = (4, -3)$ compute:

(a) $1/z_1$

(b) z_2/z_3

(c) $\text{Im}\,(z_3/z_1)$

(d) $|1/z_1|$

SOLUTION 1.2

We have:

(a) $\dfrac{1}{2+i} = \dfrac{1}{2+i}\dfrac{2-i}{2-i} = \dfrac{2-i}{5} = z_1^{-1}$

(b)

$$\frac{-1+5i}{4-3i} = \frac{-1+5i}{4-3i}\frac{4+3i}{4+3i} = \frac{-4+17i+15i^2}{16+9} = \frac{-19+17i}{25} = z_2/z_3$$

(c) $\dfrac{4-3i}{2+i} = \dfrac{4-3i}{2+i}\dfrac{2-i}{2-i} = \dfrac{5-10i}{5} = 1-2i$ and $\text{Im}\,(z_3/z_1) = -2$

(d) $|z_1^{-1}| = \sqrt{(2/5)^2 + (-1/5)^2} = 1/(\sqrt{5}) = |z_1|^{-1}$

PROBLEM 1.3

Write the following complex numbers in polar form:

$$z_1 = 1 + i\sqrt{3},\ z_2 = i,\ z_3 = \frac{i}{1+i},\ z_4 = 2,\ z_5 = (1-i)^4$$

SOLUTION 1.3

To write a complex number in polar form we must determine its modulus and (the principle value of) its argument.

for z_1: $|z_1| = \sqrt{1+3} = 2$ and $\text{Arg}\, z_1 = \text{Arctan}\sqrt{3} = \pi/3$ radians

$$\text{Then } z_1 = 2e^{i\pi/3}$$

for z_2: $|z_2| = 1$ and $\text{Arg}\, z_2 = \text{Arctan}\,(1/0) = \pi/2$ radians

$$\text{Then } z_2 = e^{i\pi/2}$$

Note: When $\text{Re}\, z = 0$ then $\text{Arg}\, z$ equals plus or minus $\pi/2$ according to whether $\operatorname{Im} z$ is positive or negative respectively.

for z_3: Note first that $1+i = \sqrt{2}e^{i\pi/4}$

$$\text{Then } z_3 = 1/\sqrt{2}e^{i(\pi/2-\pi/4)} = 2^{-1/2}e^{i\pi/4}$$

for z_4: Since $|z_4| = 2$ and $\operatorname{Arg} z_4 = 0$ we have $z_4 = 2$ for the polar form.

for z_5: Note first that $1-i = \sqrt{2}e^{-i\pi/4} = 2^{1/2}e^{-i\pi/4}$.

$$\text{Then } z_5 = (1-i)^4 = 2^2e^{-i\pi} = 4e^{-i\pi} = -4.$$

PROBLEM 1.4

Write the following complex numbers in Cartesian form, $a + ib$:

$$z_1 = 8e^{i5\pi/6},\ z_2 = 2e^{i2\pi/3},\ z_3 = 2e^{-i3\pi/4}$$

SOLUTION 1.4

The definition (1.11) implies

$$e^{i5\pi/6} = \cos 5\pi/6 + i\sin 5\pi/6 = (-\sqrt{3}+i)/2$$

Then

$$z_1 = 4(-\sqrt{3}+i) = -4\sqrt{3}+4i$$

Similarly

$$e^{i2\pi/3} = \cos 2\pi/3 + i\sin 2\pi/3 = (-1+i\sqrt{3})/2$$

$$e^{-i3\pi/4} = \cos 3\pi/4 - i\sin 3\pi/4 = \sqrt{2}(-1-i)/2$$

hence

$$z_2 = -1+i\sqrt{3} \quad \text{and} \quad z_3 = \sqrt{2}(-1-i)$$

PROBLEM 1.5

Show that the polynomial equation $x^2+x+1 = 0$ has no real solutions. Then find all the complex solutions.

SOLUTION 1.5

To see that the equation has no real solutions, note that $f(x) = x^2+x+1$ has a unique positive minimum at $x = -1/2$ (i.e., $f'(-1/2) = 0$ and $f(-1/2) = 3/4$; the graph of $f(x)$ is a parabola opening upward). Now suppose $z = (a,b)$ solves

$$z^2+z+1 = (a,b)\,(a,b) + (a,b) + 1$$

$$= (a^2-b^2, 2ab) + (a,b) + (1,0) = (0,0)$$

Then $a^2 - b^2 + a + 1 = 0$ and $2ab + b = 0$. Since there are no real zeroes for $f(x)$, b cannot equal zero. Then the second equation implies $a = -1/2$. Using this in the first equation leads to $b = \pm\sqrt{3}/2$. Then the roots of the equation are the two complex numbers

$$z_{1,2} = -1/2 \pm \sqrt{3}/2 = 1e^{\pm 2\pi/3}$$

PROBLEM 1.6

Suppose the nth degree polynomial

$$P(z) = a_n z^n + a_{n-1} z^{n-1} + \ldots + a_1 z + a_0$$

has real coefficients a_m. Show that if z_0 is any root of the equation $P(z) = 0$ then $z_0{}^*$ is also a root of this equation; i.e., complex roots to real coefficient polynomial equations occur only in complex conjugate pairs.

SOLUTION 1.6

If $P(z_0) = 0$ then

$$a_n z_0^n + a_{n-1} z_0^{n-1} + \ldots + a_1 z_0 + a_0 = 0$$

and taking the conjugate of both sides of this equation leads to

$$(a_n z_0^n + a_{n-1} z_0^{n-1} + \ldots + a_1 z_0 + a_0)^* = 0^* = 0$$

Then by the results (1.3)

$$(a_n z_0^n)^* + (a_{n-1} z_0^{n-1})^* + \ldots + (a_1 z_0)^* + a_0{}^* = 0$$

and since the coefficients are all real, $a_m{}^* = a_m$ for $m = 0$ to n. Hence

$$a_n (z_0{}^*)^n + a_{n-1} (z_0{}^*)^{n-1} + \ldots + a_1 z_0{}^* + a_0{}^* = 0$$

i.e., $P(z_0{}^*) = 0$ and z^* is a zero of the polynomial. Note that this result implies that if $P(z)$ has real coefficients and the degree of P is odd, then $P(z)$ has at least one real zero.

PROBLEM 1.7

Show that for arbitrary complex numbers z_1, z_2

$$|z_1 + z_2| \le |z_1| + |z_2| \tag{1}$$

$$|z_1 - z_2| \ge \big||z_1| - |z_2|\big| \tag{2}$$

SOLUTION 1.7

By (1.4) we have

$$|z_1+z_2|^2 = (z_1+z_2)(z_1^*+z_2^*) = z_1z_1^*+z_1z_2^*+z_1^*z_2+z_2z_2^*$$
$$= |z_1|^2+|z_2|^2+z_1z_2^*+z_1^*z_2$$
$$= |z_1|^2+|z_2|^2+2\text{Re}\, z_1z_2^*$$

But $\text{Re}\, z_1z_2^* \le |z_1z_2^*| = |z_1||z_2|$ and thus

$$|z_1+z_2|^2 \le |z_1|^2+|z_2|^2+2|z_1||z_2| = (|z_1|+|z_2|)^2. \tag{3}$$

Since the modulus is always nonnegative, we can take the square root of both sides of (3) to obtain (1). To obtain (2) we proceed in a similar fashion to get

$$|z_1-z_2|^2 = |z_1|^2+|z_2|^2-2\text{Re}\, z_1z_2^* \ge |z_1|^2+|z_2|^2-2|z_1||z_2|$$

i.e.,

$$|z_1-z_2|^2 \ge (|z_1|-|z_2|)^2. \tag{4}$$

Then taking the square root of both sides of (4), noting that $|z_1|-|z_2|$ may be negative, we obtain (2).

PROBLEM 1.8

Use deMoivre's theorem to express $\cos 4\vartheta$ and $\sin 4\vartheta$ in terms of $\sin\vartheta$ and $\cos\vartheta$.

SOLUTION 1.8

DeMoivre's theorem implies

$$\cos 4\vartheta + i\sin 4\vartheta = (\cos\vartheta + i\sin\vartheta)^4$$
$$= \cos^4\vartheta + 4\cos^3\vartheta\,(i\sin\vartheta) + 6\cos^2\vartheta\,(i\sin\vartheta)^2$$
$$+ 4\cos\vartheta\,(i\sin\vartheta)^3 + (i\sin\vartheta)^4$$
$$= \cos^4\vartheta - 6\cos^2\vartheta\sin^2\vartheta + \sin^4\vartheta + i\,(4\cos^3\vartheta\sin\vartheta - 4\cos\vartheta\sin^3\vartheta)$$

where we have used the binomial theorem to expand $(\cos\vartheta + i\sin\vartheta)^4$. Then by equating the real and imaginary parts of the two sides of this expression we find

$$\cos 4\vartheta = \cos^4\vartheta - 6\cos^2\vartheta\sin^2\vartheta + \sin^4\vartheta$$

$$\sin 4\vartheta = 4\cos^3\vartheta\sin\vartheta - 4\cos\vartheta\sin^3\vartheta$$

PROBLEM 1.9

Find all the solutions to $z^5 - 1 = 0$.

SOLUTION 1.9

Write

$$z^5 = 1 = e^{i2k\pi}.$$

Then

$$z = e^{i(2k\pi/5)} \quad \text{for } k = 0, 1, 2, 3, 4;$$

i.e.,

$$z = e^0,\ e^{i2\pi/5},\ e^{i4\pi/5},\ e^{i6\pi/5},\ e^{i8\pi/5}.$$

Since

$$e^{i6\pi/5} = e^{-i4\pi/5} \quad \text{and} \quad e^{i8\pi/5} = e^{-i2\pi/5}$$

we can also express the 5 fifth roots of 1 as

$$z = 1, e^{\pm i2\pi/5}, e^{\pm i4\pi/5}$$

which shows that these roots are uniformly distributed around the circle of radius 1 in the complex plane.

PROBLEM 1.10

Let ω denote any of the nth roots of 1 (i.e., $\omega^n = 1$). Then show that for any integer m that is not a multiple of n

$$1 + \omega^m + \omega^{2m} + \ldots + \omega^{(n-1)m} = 0 \tag{1}$$

SOLUTION 1.10

Note that if m is not a multiple of n, then ω^m is not equal to 1. Now for any complex number z that is different from 1 we have for all positive integers n

$$1 + z + z^2 + \ldots + z^{n-1} = \frac{1 - z^n}{1 - z} \tag{2}$$

To see this, let

$$S = 1 + z + z^2 + \ldots + z^{n-1}$$

Then

$$zS = z + z^2 + \ldots + z^n$$

and

$$S - zS = 1 - z^n$$

This last result implies (2). Then using (2) with $z = \omega^m$ proves (1) since in this case $z^n = 1$.

PROBLEM 1.11

Describe the following sets in the complex plane:

$D_1 = \{z\colon\ |z-i| = 2\}$ $\qquad$ $D_2 = \{z\colon\ |z-1| > 4\}$

$D_3 = \{z\colon\ 1 < |z-1| < 5\}$ $\qquad$ $D_4 = \{z\colon\ \operatorname{Re} z = 1\}$

$D_5 = \{z\colon\ \operatorname{Re} z > 1\}$ $\qquad$ $D_6 = \{z\colon\ 0 < \arg z < \pi/2\}$

SOLUTION 1.11

The set D_1 consists of all complex numbers z whose distance from the point $z_0 = i$ is equal to 2. Then D_1 is a circle in the complex plane. The center of the circle is at $z_0 = i$ and the radius is equal to 2. Note that the points that lie inside the circle satisfy $|z-i| < 2$ while those that lie outside the circle satisfy $|z-i| > 2$. The set D_1 is bounded.

The set D_2 contains all the complex numbers z whose distance from $z_0 = 1$ is *greater than* 4. Then this is the set of all z lying *outside of* the disc of radius 4 whose center is at $z_0 = 1$. D_2 is not bounded.

The set D_3 consists of the points that lie outside the disc of radius 1 with center at $z_0 = 1$ but lie inside the disc of radius 5 and center at $z_0 = 1$. This is a disc with a circular hole removed from the center. Such a region is called an *annular region*. In this case D_3 is the annular region contained between the two circles of radius 1 and 5 with centers at $z_0 = 1$. D_3 is bounded.

The complex numbers of the form $z = 1 + ib$ for all real b comprise the set D_4. This is a line in the complex plane parallel to the imaginary axis and passing through the point $z = 1$. D_4 is not bounded.

The set D_5 consists of complex numbers of the form $z = a + ib$ with $a > 1$. Then D_5 contains all points in the complex plane lying to the right of the line D_4. Note that only points to the right of D_4 are in D_5; D_4 itself does not belong to D_5. D_5 is an unbounded set.

The complex numbers z whose argument lies between 0 and $\pi/2$ are those numbers in the first quadrant of the complex plane. Then the set D_6 is this first quadrant of the plane. Note that the positive real and imaginary axes are not part of the unbounded set D_6.

PROBLEM 1.12

For each of the following sets, tell which points are interior points and which are boundary points, tell if the set is open/closed and if the set is connected.

$D_1 = \{z\colon\ |\operatorname{Re} z| \le 1 \text{ and } |\operatorname{Im} z| < 1\}$

$D_2 = \{z\colon\ \operatorname{Arg} z = \pi/4\}$

$D_1 = \{z\colon\ |\operatorname{Re} z| > 1\}$ $\quad$ and $\quad$ $D_4 =$ the whole complex plane

SOLUTION 1.12

The set D_1 is a square of sidelength 2 whose sides are parallel to the axes and whose center is at the origin. The points that are inside the square (i.e., not on one of the sides) are interior points. Thus all points z such that $|\text{Re}\, z| < 1$ and $|\text{Im}\, z| < 1$ are interior points of the set D_1. Every neighborhood of any point on one of the sides contains points that are inside the square as well as points from outside the square. Then all the points on the sides of the square are boundary points. The vertical sides of the square belong to D_1 but the horizontal sides at the top and bottom are not in D_1. Thus D_1 is not open because it contains some points (the vertical sides) that are not interior points and D_1 is not closed because it does not contain all of its boundary points (the horizontal sides are not in D_1). D_1 is connected since any two points in the square can be joined by a straight line lying inside the square.

The set D_2 is a ray that originates at the origin and makes an angle of $\pi/4$ radians with the real axis. This set has no interior points since every one of its points is a boundary point. Then D_2 is closed and it is clearly connected.

The set D_3 consists of the complex plane with a strip of width 2 removed. The centerline of the strip lies on the imaginary axis. The boundary points of D_3 are the two lines $\text{Re}\, z = 1$ and $\text{Re}\, z = -1$. None of these points belong to D_3 so D_3 is not closed. It is easy to see that all points of D_3 are interior points, hence D_3 is open. It is also clear that D_3 is not connected since any point z_1 with $\text{Re}\, z_1 > 1$ cannot be joined to a point z_2 with $\text{Re}\, z_2 < -1$ by a polygonal path that remains in D_3.

Every neighborhood of any point in D_4 is contained in D_4 since D_4 is the whole plane. Thus every point of D_4 is an interior point and therefore D_4 is open. Since all points belong to D_4, there are no boundary points of D_4 that are not in D_4. Thus D_4 is closed. Finally, D_4 is connected since every polygonal path lies within D_4.

SUMMARY

Complex numbers are defined to be ordered pairs of real numbers $z = (a, b)$ with sums and products defined by

$$z_1 + z_2 = (a_1 + a_2, b_1 + b_2)$$

$$z_1 z_2 = (a_1 a_2 - b_1 b_2, a_1 b_2 + a_2 b_1) .$$

If we define the complex conjugate of $z = (a, b)$ to be the number $z^* = (a, -b)$ and define modulus of z by $|z| = (zz^*)^{1/2} = (a^2 + b^2)^{1/2}$ then the quotient of two complex numbers is given by

$$\frac{z_1}{z_2} = \frac{z_1 z_2^*}{z_2 z_2^*} = \frac{1}{|z_2^*|^2} z_1 z_2^* = \left(\frac{a_1 a_2 + b_1 b_2}{|z_2|^2}, \frac{a_2 b_1 - a_1 b_2}{|z_2|^2} \right)$$

We have the following alternative representations for the complex number $z = (a, b)$:

Cartesian $\quad z = a + ib \quad$ where $\quad i^2 = -1$

Polar $\quad z = re^{i\vartheta} \quad$ where $\quad r = |z| \quad \tan\vartheta = b/a$

$$e^{i\vartheta} = \cos\vartheta + i\sin\vartheta$$

In polar notation

$$z_1 z_2 = r_1 r_2 e^{i(\vartheta_1 + \vartheta_2)} \quad \text{and} \quad z_1/z_2 = r_1/r_2 e^{i(\vartheta_1 - \vartheta_2)}$$

$$z^n = r^n e^{in\vartheta} \quad \text{and} \quad z^{1/n} = r^{1/n} e^{i(\vartheta + 2\pi k)/n} \quad \text{for } k = 0 \text{ to } n - 1.$$

We will show later that every polynomial equation of degree n has n complex roots. We have shown already that if the polynomial equation has real coefficients then the complex roots can only occur in complex conjugate pairs.

SUPPLEMENTARY PROBLEMS

1. Write each of the following complex numbers in polar form:

   $z_1 = 1 + i\sqrt{3},\ z_2 = 2 + 2i,\ z_3 = -\sqrt{3} + i,\ z_4 = 1 - i$

2. Compute $z_1 + z_2,\ z_2 z_3,\ z_3/z_4$

3. Write each of the following complex numbers in Cartesian form:

   $w_1 = 2e^{i3\pi/4},\ w_2 = e^{-i\pi/6},\ w_3 = 6e^{i5\pi/6},\ w_4 = 4e^{-i2\pi/3}$

4. Compute $w_1 + w_2,\ w_2 w_3,\ w_3/w_4$.

5. Compute:

   (a) $\dfrac{(3+i)(2-i)}{1+i}$

(b) $\dfrac{(3-2i)^2}{|3-2i|}$

(c) $\dfrac{i}{(1-i)(2+i)}$

6. Compute:

(a) $\text{Im}\,\dfrac{3+i}{-1+2i}$

(b) $\dfrac{|21-55i|^2}{21+55i}$

7. Find all the solutions to $z^4 + 16i = 0$.

8. Find all the solutions to $z^3 = 1 + i\sqrt{3}$.

9. Describe the set of points z satisfying:
(a) $|z-1| - |z+1| > 2$
(b) $|z-1| - |z+1| = 2$

SOLUTIONS TO SUPPLEMENTARY PROBLEMS

1. $z_1 = 2e^{i\pi/3}$, $z_2 = 2\sqrt{2}e^{i\pi/4}$, $z_3 = 2e^{i5\pi/6}$, $z_4 = \sqrt{2}e^{-i\pi/4}$

2. $z_1 + z_2 = 3 + i(2+\sqrt{3})$, $z_2 z_3 = -2 - 2\sqrt{3} + i(2 - 2\sqrt{3})$,

$z_3/z_4 = (-1-\sqrt{3}+i(1-\sqrt{3}))/2$

3. $w_1 = \sqrt{2}(-1+i)$, $w_2 = (\sqrt{3}-i)/2$, $w_3 = 3(-\sqrt{3}+i)$,

$w_4 = -2(1+i\sqrt{3})$

4. $w_1 + w_2 = (\sqrt{3}-2\sqrt{2})/2 + i(2\sqrt{2}-1)/2$, $w_2 w_3 = 6e^{i2\pi/3}$,

$w_3/w_4 = -\dfrac{3}{2}i$

5. (a) $3-4i$

(b) $(5-12i)/\sqrt{13}$

(c) $(-1+3i)/10$

6. (a) $-7/5$
   (b) $21-55i$

7. $z = 2e^{-i\pi/8},\ 2e^{i3\pi/8},\ 2e^{i7\pi/8},\ 2e^{i11\pi/8}$

8. $z = 2^{1/3}e^{i\pi/9},\ 2^{1/3}e^{i7\pi/9},\ 2^{1/3}e^{i13\pi/9}$

9. (a) The triangle inequality shows there are no z satisfying this condition.
   (b) The condition is satisfied for all z on the real axis, $z = (x, 0)$, with $x \leq -1$.

2

Functions of a Complex Variable: Differentiation

This chapter and the next two contain the elements of what is generally referred to as classical complex function theory. In this chapter we introduce the notion of a complex valued function of a complex variable. As is usually done for real valued functions, we define the concept of a limit and use this to introduce continuity and differentiability for complex valued functions.

Discussions of differentiation bring to light the important Cauchy Riemann equations and lead to the related notions of analytic and harmonic functions. The chapter concludes with an examination of the so-called elementary functions (polynomials, rational functions, the exponential and log functions and the trigonometric functions) as complex valued functions of a complex variable.

COMPLEX VALUED FUNCTIONS OF A COMPLEX VARIABLE

NOTATION

Let f denote a complex valued function of the complex variable z; i.e., $w = f(z)$ where $z = x+iy$ and $w = u+iv$. Then $f(x+iy) = u(x,y) + iv(x,y)$ and we may view $u = \text{Re} f$ and $v = \text{Im} f$ as two real valued functions of the two real variables x and y. Alternatively we have $w = f(z)$ where $z = re^{i\vartheta}$ and $w = \rho e^{i\vartheta}$. Then $\rho = |f|$ and $\varphi = \text{Arg} f$ are two real valued functions of the two real variables r and ϑ.

EXAMPLE 2.1

(a) Consider the function $f(z) = z^2$. Then

$$u + iv = (x + iy)^2 = x^2 - y^2 + i2xy;$$

i.e.,

$$u(x, y) = x^2 - y^2 \quad \text{and} \quad v(x, y) = 2xy.$$

Alternatively

$$\rho e^{i\varphi} = (re^{i\vartheta})^2 = r^2 e^{i2\vartheta};$$

i.e.,

$$\rho = r^2 \quad \text{and} \quad \varphi = 2\vartheta$$

(b) For $f(z) = zz^* = |z|^2 = x^2 + y^2$ we have

$$u + iv = x^2 + y^2 \text{ hence } u(x, y) = x^2 + y^2 \text{ and } v(x, y) = 0$$

Also $\rho e^{i\varphi} = r^2$ hence $\rho = r^2$ and $\varphi = 0$.

(c) For $f(z) = 1/z$ for z not zero we have

$$f(z) = \frac{1}{z} = \frac{z^*}{zz^*} = \frac{x}{x^2 + y^2} - i\frac{y}{x^2 + y^2}$$

thus

$$u(x, y) = \frac{x}{x^2 + y^2} \quad \text{and} \quad v(x, y) = \frac{-y}{x^2 + y^2}.$$

In polar notation

$$f(re^{i\vartheta}) = r^{-1}e^{-i\vartheta}$$

hence

$$\rho = r^{-1} \quad \text{and} \quad \varphi = -\vartheta.$$

Complex Limits

The complex valued function $f(z)$ is said to have *limit* L at $z = z_0$ if and only if for every $\varepsilon > 0$ there exists a $\delta > 0$ such that $|f(z) - L| < \varepsilon$ for all z such that $0 < |z - z_0| < \delta$. Note that the definition says that the limit exists and equals L if $f(z)$ is close to L whenever z is sufficiently close to, *but not equal to*, z_0. Complex limits are very similar to limits of real valued functions of two real variables. In particular, the limiting value of $f(z)$ must be independent of the path along which the limit point z_0 is approached.

Theorem 2.1

Suppose that $f(z)$ tends to distinct values L_1 and L_2 as z approaches z_0 along different paths. Then the limit of $f(z)$ as z tends to z_0 does not exist.

Theorem 2.2

The limit of $f(z) = u(x, y) + iv(x, y)$ as z tends to $z_0 = x_0 + iy_0$ exists and equals $L = u_0 + iv_0$ if and only if the limits of the real valued functions $u(x, y)$ and $v(x, y)$ as (x, y) tends to (x_0, y_0) both exist and equal u_0 and v_0 respectively.

Theorem 2.3

Suppose the limits of the complex valued functions $f(z)$ and $g(z)$ as z tends to z_0 both exist and equal L_0 and M_0 respectively. Then each of the following limits exists and has the indicated value:

1. $\lim_{z \to z_0} C_1 f(z) + C_2 g(z) = C_1 L_0 + C_2 M_0$ for all complex constants C_1, C_2.
2. $\lim_{z \to z_0} f(z) g(z) = L_0 M_0$
3. $\lim_{z \to z_0} f(z) / g(z) = L_0 / M_0$ provided M_0 is not zero.
4. $\lim_{z \to z_0} f(z)^* = L_0^*$, $\lim_{z \to z_0} \text{Re} f(z) = \text{Re} L_0$, $\lim_{z \to z_0} \text{Im} f(z) = \text{Im} L_0$

Continuity

The complex valued function $f(z)$ is said to *continuous at* $z = z_0$ if and only if the limit of $f(z)$ as z tends to z_0 exists and equals $f(z_0)$. Note that this definition requires f to be defined at z_0, it requires f to tend to a limit at z_0 and finally it requires the limit to equal the function value. Equivalently we can define f to be continuous at $z = z_0$ if and only if for every $\varepsilon > 0$ there exists a $\delta > 0$ such that $|f(z) - f(z_0)| < \varepsilon$ whenever $|z - z_0| < \delta$. The function f is said to be continuous on the domain D if $f(z)$ is continuous at each point z of D. The continuity of $f(z)$ is equivalent to the continuity of the real valued functions $\text{Re} f$ and $\text{Im} f$.

Theorem 2.4

The complex valued function $f(z) = u(x, y) + iv(x, y)$ is continuous at $z = z_0 = x_0 + iy_0$ if and only if $u = u(x, y)$ and $v = v(x, y)$ are both continuous at (x_0, y_0). The function $f(z)$ is continuous for all z in D if and only if $u(x, y)$ and $v(x, y)$ are continuous for all (x, y) in D. In this case if D is closed and bounded then the functions $f(z)$, $u(x, y)$ and $v(x, y)$ are all uniformly continuous on D, hence there exist constants K, L and M such that $|f(z)| \leq M$ for all z in D and $|u(x, y)| \leq K$, $|v(x, y)| \leq L$ for all (x,y) in D.

Theorem 2.5 Suppose $f(z)$ and $g(z)$ are both continuous at $z = z_0$. Then each of the following is also continuous at z_0:

1. Linear Combinations: $C_1 f(z) + C_2 g(z)$ for all complex constants C_1, C_2
2. Product: $f(z)\, g(z)$
3. Quotient: $f(z)/g(z)$ as long as $g(z_0)$ is not zero
4. Composition: If $g(z)$ is continuous at $z = z_0$ and $f(w)$ is continuous at $w_0 = g(z_0)$ then $f(g(z))$ is continuous at z_0.

Differentiation

Let the complex valued function f of the complex variable z be defined and continuous in a neighborhood of z_0. Then we say $f(z)$ is *differentiable* at z_0 with derivative $f'(z_0)$ given by

$$f'(z_0) = \lim_{z \to z_0} \frac{f(z) - f(z_0)}{z - z_0} \tag{2.1}$$

if the limit exists. While the definition of the derivative for a complex valued function is superficially the same as the definition of derivative for a real valued function, the existence of the complex derivative implies something more about the function f than is implied by the existence of the derivative for a real valued function.

EXAMPLE 2.2

(a) Consider the function $f(z) = z^2$. For this function the difference quotient in (2.1) assumes the form

$$\frac{\Delta f}{\Delta z} = \frac{z^2 - z_0^2}{z - z_0} = z + z_0.$$

As z tends to z_0 this expression tends to the limit $2z_0$ independent of the path along which z approaches z_0. Thus $f'(z_0)$ exists and equals $2z_0$ at all points z_0. We can use similar constructions to find the derivatives for other elementary functions of z.

(b) Suppose the function $f(z)$ is a *real valued* function of the complex variable z defined in a neighborhood of z_0. Suppose further that the limit (2.1) exists. Then Theorem 2.1 implies that for any real h tending to zero, the expressions

$$\frac{f(z_0 + h) - f(z_0)}{h} \quad \text{and} \quad \frac{f(z_0 + ih) - f(z_0)}{ih}$$

must both tend to the limiting value $f'(z_0)$. Since the numerator in both of the above quotients is real, the first expression must converge to a real limit and the second must converge to an imaginary limiting value. But these two limiting values are both equal to $f'(z_0)$ which must therefore be equal to zero. This proves that if a real valued function of a complex variable is differentiable at a point then the derivative at the point must vanish.

DIFFERENTIATION FORMULAS

While the derivative of a complex valued function of a complex variable will be seen to differ in some ways from the derivative of a real valued function of a single real variable, the differentiation formulas for the elementary functions of z are formally identical to the formulas for the derivatives of real valued elementary functions. For example, by means like those used in Example 2.2(a) the derivative of $f(z) = z^n$ is found to equal $f'(z) = nz^{n-1}$. Similarly, all of the differentiation rules such as the product, quotient and chain rules apply to complex valued functions of a complex variable precisely as they apply in the real case.

THE CAUCHY RIEMANN EQUATIONS

As we shall now see, the existence of the derivative of a complex valued function of a complex variable implies a certain linkage between the derivatives of the real and imaginary parts of the function. This linkage later turns out to have far reaching consequences.

Theorem 2.6 Suppose $f(z) = u(x, y) + iv(x, y)$ is defined in a neighborhood of the point z_0.

1. If the limit (2.1) exists then the partial derivatives $\partial_x u$, $\partial_y u$, $\partial_x v$ and $\partial_y v$ all exist at (x_0, y_0) and at that point we must have

$$\partial_x u = \partial_y v \quad \text{and} \quad \partial_y u = -\partial_x v \tag{2.2}$$

2. If u and v are continuously differentiable throughout the neighborhood of $z_0 = (x_0, y_0)$ and if the equations (2.2) are satisfied at (x_0, y_0) then f is differentiable at z_0 and

$$\begin{aligned} f'(z_0) &= \partial_x u(x_0, y_0) + i\partial_x v(x_0, y_0) \\ &= \partial_y v(x_0, y_0) - i\partial_y u(x_0, y_0) \end{aligned} \tag{2.3}$$

The equations (2.2) are called the *Cauchy Riemann equations.*

Sometimes it is more convenient to state the results of the previous theorem in polar notation.

Theorem 2.7 Suppose $f(z) = u(r, \vartheta) + iv(r, \vartheta)$ is defined in a neighborhood of the point z_0.

1. If the limit (2.1) exists then the partial derivatives $\partial_r u$, $\partial_\vartheta u$, $\partial_r v$ and $\partial_\vartheta v$ all exist at (r_0, ϑ_0) and at that point we must have

$$r\partial_r u = \partial_\vartheta v \text{ and } \partial_\vartheta u = -r\partial_r v \tag{2.4}$$

2. If u and v are continuously differentiable throughout the neighborhood of $z_0 = (r_0, \vartheta_0)$ and if the equations (2.4) are satisfied at (r_0, ϑ_0) then f is differentiable at z_0 and

$$\begin{aligned} f'(z_0) &= e^{-i\vartheta_0}(\partial_r u(r_0, \vartheta_0) + i\partial_r v(r_0, \vartheta_0)) \\ &= \frac{e^{-i\vartheta_0}}{r_0}(\partial_\vartheta v(r_0, \vartheta_0) - i\partial_\vartheta u(r_0, \vartheta_0)) \end{aligned} \tag{2.5}$$

The equations (2.4) are the Cauchy Riemann equations expressed in polar form.

ANALYTIC FUNCTIONS

If $f'(z)$ exists at $z = z_0$ and throughout a neighborhood of z_0 as well, then f is said to be *analytic* at z_0. The word *holomorphic* is sometimes used in place of analytic. If $f = (z)$ is analytic at each point of a domain D then f is said to be analytic on D. If $f(z)$ is analytic at every point in the complex plane then f is said to be an *entire function*. Note that in order for f to be analytic at z_0 it is not sufficient for the Cauchy Riemann equations to be satisfied just at the single point z_0 nor even is it enough for them to hold along a line or a curve through z_0; the equations must be satisfied throughout a neighborhood $N_r(z_0)$ for some $r > 0$.

EXAMPLE 2.3

(a) For $f(z) = z^2 = x^2 - y^2 + i2xy$ we compute

$$\partial_x u = 2x,\ \partial_y u = -2y \quad \text{and} \quad \partial_x v = 2y,\ \partial_y v = 2x.$$

The functions $u(x, y)$ and $v(x, y)$ are continuously differentiable and the Cauchy Riemann equations are satisfied at each point (x, y) in the plane. The $f(z) = z^2$ is analytic in the whole complex plane by Theorem 2.6; i.e., $f(z) = z^2$ is an entire function. It is easy to show that every polynomial in z is entire.

(b) For $f(z) = zz^* = x^2 + y^2$ we find

$$\partial_x u = 2x,\ \partial_y u = 2y \quad \text{and} \quad \partial_x v = \partial_y v = 0.$$

Then the functions u and v are everywhere smooth but the Cauchy Riemann equations are satisfied only at the origin. Then $f'(0)$ exists and equals zero but f is not analytic at the origin since $f'(z)$ does not exist throughout any neighborhood of $z = 0$.

Theorem 2.8

If f and g are analytic in domain D then linear combinations of f and g are analytic in D as is the product fg. The quotient f/g is analytic on D if g does not vanish at any point of D. If F is a function analytic on a domain containing the range of f then the composed function $F(w) = F(f(z))$ is analytic on D.

Theorem 2.9

If $f(z)$ is analytic at z_0 then $f(z)$ is continuous at z_0.

HARMONIC FUNCTIONS

The real valued function $u = u(x, y)$ is said to be *harmonic* in D if u is twice continuously differentiable with respect to x and y in D and if $u(x, y)$ solves *Laplace's equation* $\partial_{xx}u(x, y) + \partial_{yy}u(x, y) = 0$ throughout D. If u and v are harmonic in D and also satisfy the Cauchy Riemann equations in D then u and v are said to be *harmonic conjugates.*

ANALYTICITY AND HARMONIC FUNCTIONS

Supose $f(z) = u(x, y) + iv(x, y)$ is analytic in a region D. Then $\partial_x u$, $\partial_y u$, $\partial_x v$ and $\partial_y v$ all exist and satisfy the Cauchy Riemann equations throughout D. It will be shown later that as the real and imaginary parts of an analytic function, the real valued functions $u(x, y)$ and $v(x, y)$ necessarily have derivatives of all orders at every point in D. For all smooth functions $u = u(x, y)$ and $v = v(x, y)$ we have $\partial_x(\partial_y u) - \partial_y(\partial_x u) = 0$ and $\partial_x(\partial_y v) - \partial_y(\partial_x v) = 0$ at each point of D. For $u = \operatorname{Re} f$ and $v = \operatorname{Im} f$, in particular, it follows from the Cauchy Riemann equations that $\partial_{xx}u(x, y) + \partial_{yy}u(x, y) = 0$ and $\partial_{xx}v(x, y) + \partial_{yy}v(x, y) = 0$ in D.

Theorem 2.10

If $f = f(z)$ is analytic in the domain D then $u = \operatorname{Re} f$ and $v = \operatorname{Im} f$ are harmonic conjugates of one another throughout D.

EXAMPLE 2.4

Consider the function $u(x, y) = x^3 - 3xy^2$. We may show in two ways that this is a harmonic function on the entire complex plane. First,

$$\partial_x u = 3x^2 - 3y^2 \quad \text{and} \quad \partial_{xx} u = 6x$$

$$\partial_y u = -6xy \quad \text{and} \quad \partial_{yy} u = -6x$$

from which it is evident that u and its derivatives are everywhere continuous and $\partial_{xx} u + \partial_{yy} u = 0$ at every point in the complex plane. A second proof that u is harmonic on the entire plane makes use of Theorem 2.10. We observe that

$$f(z) = z^3 = (x+iy)^3 = x^3 - 3xy^2 + i(3x^2 y - y^3)$$

$$= u(x, y) + iv(x, y).$$

Thus $u(x, y)$ is the real part of a function which can be easily seen to be entire. It is furthermore clear from this observation that $v(x, y) = 3x^2 y - y^3$ is also harmonic and is, in fact, the harmonic conjugate of $u(x, y)$. See Problem 2.9 for a means of finding v directly from u without recourse to the complex function $f(z)$.

ORTHOGONAL TRAJECTORIES

The level curves of a harmonic function u can be shown to be an orthogonal family to the level curves of the harmonic conjugate for u, (see Problem 2.10). The two families of orthogonal curves are referred to as *orthogonal trajectories*.

Elementary Functions

POLYNOMIALS

The function $f(z) = C$ (C = complex constant) is an entire function with derivative everywhere equal to zero. The function $f(z) = z$ is easily shown to be entire with derivative equal to the constant 1. Since linear combinations and products of analytic functions are again analytic by Theorem 2.8, it follows that the nth degree polynomial

$$P_n(z) = z^n + a_{n-1} z^{n-1} + \ldots + a_1 z + a_0$$

with complex coefficients a_m is an entire function whose derivative is equal to

$$P_n{}'(z) = nz^{n-1} + (n-1)a_{n-1} z^{n-2} + \ldots + a_1.$$

The fundamental theorem of algebra, to be proved later, asserts that for $n > 0$, $P_n(z) = 0$ has at least one solution in the set of complex numbers. If $P_n(z_1) = 0$ then it follows that $P_n(z) = (z - z_1) P_{n-1}(z)$

for P_{n-1} a polynomial of degree $n-1$. Successive applications of the fundamental theorem of algebra to $P_{n-1}, P_{n-2}, \ldots$ leads to the result

$$P_n(z) = (z-z_1)^a (z-z_2)^b \ldots (z-z_m)^p \text{ where } a+b+\ldots+p = n;$$

i.e., $P_n(z)$ has n (not necessarily distinct) zeroes, z_1 to z_m having *multiplicities* $a, b, \ldots, p$ respectively. Note that if z_j is a root of multiplicity r for $P_n(z) = 0$, then $P_n(z) = (z-z_j)^r P_{n-r}(z)$ with $P_{n-r}(z_j)$ not equal to zero. In addition,

$$P_n^{(m)}(z_j) = 0 \text{ for } m = 0 \text{ to } r-1 \quad \text{and} \quad P_n^{(r)}(z_j) \neq 0.$$

RATIONAL FUNCTIONS

A function of the form $R(z) = P_m(z)/Q_n(z)$ where P_m and Q_n are polynomials of degree m and n respectively is called a *rational function* of z. We suppose P_m and Q_n have no common factors. The function $R(z)$ is analytic everywhere except at the zeroes of $Q_n(z)$. Since $R(z)$ is analytic at all points in any neighborhood of each of these zeroes, we say that the zeroes of $Q_n(z)$ are *isolated singularities* of $R(z)$, or *pole type* singularities. If $z = z_j$ is a root of multiplicity p for $Q_n(z) = 0$ then we say z_j is a *pole of order* p for $R(z)$. If $p = 1$ we say z_j is a *simple pole.*

PARTIAL FRACTION DECOMPOSITION

If $Q_n(z) = q_n(z-z_1)\ldots(z-z_n)$ has n distinct zeroes and degree $P_m = m < n$ then

$$R(z) = \frac{A_1}{z-z_1} + \ldots + \frac{A_n}{z-z_n}$$

where $A_k = P_m(z_k)/Q'_n(z_k)$, for $k = 1$ to n. Similar partial fraction expansions for $R(z)$ are possible in case $Q_n(z)$ has multiple zeroes.

THE EXPONENTIAL FUNCTION

In the previous chapter we stated Euler's formula as a definition and observed that the definition is consistent with the usual rules for real exponents. Now we will define the exponential function $f(z) = e^z$ and show that Euler's formula follows from applying the definition with $z = iy$. We define $f(z) = e^z$ to have three properties:

i) $f(z)$ is everywhere analytic

ii) $f'(z) = f(z)$ for all z

iii) $f(z) = e^x$ if $z = x$ (i.e., $f(z)$ reduces to the usual exponential function if z is real)

In the solved problems we show that these conditions imply

$$f(z) = e^z = e^x(\cos y + i \sin y) \tag{2.6}$$

In particular, Euler's formula follows from (2.6) with $z = iy$.

TRIGONOMETRIC FUNCTIONS

For real ϑ, it follows from (2.6) that

$$\sin\vartheta = (e^{i\vartheta} - e^{-i\vartheta})/2i \quad \text{and} \quad \cos\vartheta = (e^{i\vartheta} + e^{-i\vartheta})/2$$

This motivates the definition for complex z

$$\sin z = (e^{iz} - e^{-iz})/2i \quad \text{and} \quad \cos z = (e^{iz} + e^{-iz})/2 \tag{2.7}$$

It is clear from (2.7) that $\sin z$ and $\cos z$ are entire functions. Equivalently, we can use the trigonometric identities for the sine and cosine of the sum of two arguments to obtain

$$\begin{aligned} \sin z &= \sin(x+iy) = \sin x\cosh y + i\cos x\sinh y \\ \cos z &= \cos(x+iy) = \cos x\cosh y - i\sin x\sinh y \end{aligned} \tag{2.8}$$

The remaining complex trigonometric functions can now be defined from the complex sine and cosine functions.

THE COMPLEX LOGARITHM

We define the complex logarithm by

$$w = \log z \text{ if and only if } z = e^w \text{ (i.e., } re^{i\vartheta} = e^{u+iv}) \tag{2.9}$$

Then

$$\log z = u + iv = \ln r + i\vartheta \qquad -\pi < \vartheta \le \pi \tag{2.10}$$

MULTIPLE VALUED FUNCTIONS

The polar coordinates of a point z in the complex plane are not unique. For each $z = re^{i\vartheta}$ we have

$$z = z_k = re^{i(\vartheta + 2k\pi)} \quad \text{for any integer } k.$$

This causes some functions to have more than one value at each z.

EXAMPLE 2.5

(a) The square root function $f(z) = \sqrt{z} = r^{1/2}e^{i(\vartheta+2k\pi)/2}$ has two distinct values at each z

$$w_o = r^{1/2}e^{i\vartheta/2} \quad \text{if } k \text{ is even}$$

and

$$w_1 = r^{1/2}e^{i\vartheta/2}e^{i\pi} = -r^{1/2}e^{i\vartheta/2} \quad \text{if } k \text{ is odd.}$$

Then $f(z) = z^{1/2}$ is said to be a *double valued function.*

(b) The function $F(z) = \log z$ has an infinity of values at each z

$$F(z) = \log z = \ln r + i(\vartheta + 2k\pi) \qquad k = \text{integer}$$

Any function that assumes more than a single value at each z is said to be *multiple valued*.

BRANCHES OF MULTIPLE VALUED FUNCTIONS

The functions $f(z) = z^{1/2}$ and $F(z) = \log z$ are each single valued when restricted to the domain

$$D_k = \{z = re^{i\vartheta}: \ r>0,\ \alpha + 2k\pi < \vartheta \le \alpha + 2(k+1)\pi\}$$

for α a fixed real number and k a fixed integer. The values of f and F on D_k are not the same as the values obtained for f and F on D_{k+1}. We refer to these as distinct *branches* of the functions f and F. The function $f(z) = z^{1/2}$ has only two distinct branches while $F(z) = \log z$ has infinitely many different branches. The branches obtained by choosing $\alpha = -\pi$ and $k = 0$ are called the *principal branches* for each of these functions. Note that

$$(\sqrt{z})^2 = z \quad \text{and} \quad e^{\log z} = z \quad \text{for all branches of } f \text{ and } F$$

but $$\sqrt{z^2} = re^{i(\vartheta + 2k\pi)} \quad \text{and} \quad \log(e^z) = \ln e^x + i(y + 2k\pi)$$

and these last two expressions equal z only for the principal branches of each function.

BRANCH CUTS

The ray $\vartheta = \alpha$ in the complex plane is called the *branch cut* for the branches of the functions f and F. The branch cut is a barrier separating the domain of one branch of a function from the domain of a different branch. A multiple valued function is discontinuous at every point of a branch cut for the function (see Problem 2.2). Each branch of the functions $\sqrt{z}$ and $\log z$ is analytic on its domain but is not analytic at any point of the branch cut bounding the domain.

SOLVED PROBLEMS

Limits

PROBLEM 2.1

Show that $f(z) = z^2$ tends to the limit $L = -1$ as z approaches the value $z_0 = i$

SOLUTION 2.1

Note that

$$|f(z)-L|^2 = (x^2-y^2+1)^2 + (2xy)^2 \tag{1}$$

and

$$|z-z_0|^2 = |(x,y)-(0,1)|^2 = x^2+(y-1)^2. \tag{2}$$

Since (x, y) is tending to the value $(0, 1)$ we are entitled to confine our attention to (x, y) such that $-1 < x < 1$ and $0 < y < 2$. Then (1) implies

$$\begin{aligned} |f(z)-L|^2 &= x^4 - 2x^2(y^2-1) + (y^2-1)^2 + (2xy)^2 \\ &= x^4 + 2x^2(y^2+1) + (y^2-1)^2 \\ &= x^2(x^2+2(y^2+1)) + (y+1)^2(y-1)^2 \end{aligned}$$

i.e.,

$$|f(z)-L|^2 \le 16(x^2+(y-1)^2) \tag{3}$$

for $-1 < x < 1$ and $0 < y < 2$. Now for given $\varepsilon > 0$ let $\delta = \varepsilon/4$. If $|z-z_0| < \delta$ then it follows from (2) and (3) that

$$|f(z)-L|^2 \le 16(x^2+(y-1)^2) < 16\delta^2 = \varepsilon \tag{4}$$

Thus by definition, the limit exists and equals $L = -1$.

PROBLEM 2.2

Use Theorem 2.1 to show that the limit of $f(z) = \sqrt{z}$ as z tends to -1 fails to exist.

SOLUTION 2.2

We begin by writing

$$f(z) = z^{1/2} = r^{1/2}e^{i\vartheta/2}$$

and

$$z_0 = -1 = 1e^{i\pi}.$$

Now let $\xi_n = \cos\pi(1-1/n) + i\sin\pi(1-1/n)$ for $n = 1, 2, \ldots$ and note that $|\xi_n - z_0|$ can be made arbitrarily small by choosing n large and that as n increases

$$f(\xi_n) = \cos(\pi(1-1/n)/2) + i\sin(\pi(1-1/n)/2)$$

approaches the value i. On the other hand for

$$\alpha_n = \cos\pi(-1-1/n) + i\sin\pi(-1-1/n) \quad \text{for } n = 1, 2, \ldots$$
$$= \cos\pi(1+1/n) - i\sin\pi(1+1/n)$$

we can make $|\alpha_n - z_0|$ as small as we like by choosing n large and in this case

$$f(\alpha_n) = \cos(\pi(-1-1/n)/2) + i\sin(\pi(-1-1/n)/2)$$

approaches the value $-i$ as n increases. Thus $f(\xi_n)$ approaches the value $L_1 = i$ as we approach the point $z_0 = -1$ along a path in the upper half-plane while $f(\alpha_n)$ tends to the value $L_2 = -i$ as we approach z_0 along a path in the lower half plane. Then the theorem asserts that the limit of $f(z)$ as z tends to $z_0 = -1$ does not exist. Note that -1 lies on the branch cut for the function. The square root function is analytic at every point of the complex plane except at points of the branch cut.

Continuity

PROBLEM 2.3

Show that the function $f(z) = 1/z$ is continuous on any closed set D in the complex plane that excludes the origin.

SOLUTION 2.3

If D is a closed set excluding the origin then the origin is neither an interior point nor a boundary point of D. Then there is a neighborhood $N_a(0)$ of the origin that contains no points of D. Let E_a denote the set $\{z: (|z| \geq a)\}$; this is the complex plane with this neighborhood removed. Then D is contained in the set E_a and we can prove the result by showing that $f(z) = 1/z$ is continuous on E_a for every $a > 0$.

Write $|f(z_1) - f(z_2)| = |1/z_1 - 1/z_2| = |z_2 - z_1|/|z_1 z_2|$. Then it is clear that

$$|f(z_1) - f(z_2)| \leq |z_2 - z_1|/a^2 \quad \text{for all } z_1, z_2 \text{ in } E_a.$$

Therefore, for each $\varepsilon > 0$ we have $|f(z_1) - f(z_2)| \leq \varepsilon$ for all z_1, z_2 in E_a such that $|z_2 - z_1| < \varepsilon a^2 = \delta$. Then f is continuous on E_a for every $a > 0$.

Differentiability

PROBLEM 2.4

Show that $f(z) = |z|$ is not differentiable at any nonzero z.

SOLUTION 2.4

To begin, we form the difference quotient

$$\frac{\Delta f}{\Delta z} = \frac{|z+\Delta z| - |z|}{\Delta z} = \frac{|z+\Delta z|^2 - |z|^2}{\Delta z\,(|z+\Delta z| + |z|)}$$
$$= \frac{z^* + z\,(\Delta z^*/\Delta z)}{|z+\Delta z| + |z|}$$

Approaching z along a line parallel to the real axis, with $\Delta z = \Delta x = \Delta z^*$, we have

$$\frac{\Delta f}{\Delta z} = \frac{2x}{|z+\Delta x| + |z|} \rightarrow \frac{x}{|z|} \quad \text{as } \Delta x \rightarrow 0 \tag{1}$$

and the difference quotient tends to a real limit as Δz tends to zero. Similarly, if $\Delta z = i\Delta y = -\Delta z^*$, so that we approach z along a line parallel to the imaginary axis, we find

$$\frac{\Delta f}{\Delta z} = \frac{-i2y}{|z+i\Delta y| + |z|} \rightarrow \frac{-iy}{|z|} \quad \text{as } \Delta y \rightarrow 0. \tag{2}$$

Then the difference quotient tends to an imaginary limit and, for z not equal to zero, the two limits (1) and (2) cannot be equal. It follows that $f(z) = |z|$ is not differentiable at any nonzero z.

Note that for $f(z) = |z| = (x^2+y^2)^{1/2}$ we have

$$\partial_x u = x/(x^2+y^2)^{1/2},\ \partial_y u = y/(x^2+y^2)^{1/2} \text{ and } \partial_x v = \partial_y v = 0.$$

Then the derivatives of u and v are continuous at the origin and the Cauchy Riemann conditions are satisfied at $z = 0$ so $f'(0)$ exists and equals zero. The Cauchy Riemann conditions are not satisfied throughout any neighborhood of zero so $f(z)$ is not analytic at $z = 0$.

Cauchy Riemann Equations

PROBLEM 2.5

Suppose $f(z) = u(x, y) + iv(x, y)$ is differentiable at z_0 and $u(x, y)$, $v(x, y)$ are continuously differentiable in a neighborhood of $z_0 = (x_0, y_0)$. Then show that u and v must satisfy the Cauchy Riemann equations at the point (x_0, y_0).

SOLUTION 2.5

If $f(z)$ is differentiable at $z = z_0$ then the limit (2.1) exists; i.e., the difference quotient

$$\frac{f(z) - f(z_0)}{z - z_0} = \frac{u(x, y) - u(x_0, y_0) + i\,(v(x, y) - v(x_0, y_0))}{x - x_0 + i\,(y - y_0)} \tag{1}$$

tends to a limiting value $f'(z_0)$ as z tends to z_0. If z_0 is approached along a straight line path parallel to the real axis then (x, y_0) tends to (x_0, y_0) and (1) reduces to

$$\frac{f(z) - f(z_0)}{z - z_0} = \frac{u(x, y) - u(x_0, y_0) + i(v(x, y) - v(x_0, y_0))}{x - x_0}$$

and letting z tend to z_0 along this path leads to

$$\lim_{z \to z_0} \frac{f(z) - f(z_0)}{z - z_0} = \partial_x u(x_0, y_0) + i\partial_x v(x_0, y_0) \tag{2}$$

On the other hand, letting z tend to z_0 along a path that is parallel to the imaginary axis reduces (1) to the expression

$$\frac{f(z) - f(z_0)}{z - z_0} = \frac{u(x_0, y) - u(x_0, y_0) + i(v(x_0, y) - v(x_0, y_0))}{i(y - y_0)}$$

Then as y approaches y_0 we obtain

$$\lim_{z \to z_0} \frac{f(z) - f(z_0)}{z - z_0} = -i\partial_y u(x_0, y_0) + \partial_y v(x_0, y_0) \tag{3}$$

But the fact that the limit exists implies that the limiting values in (2) and (3) must be equal. That is,

$$\partial_x u(x_0, y_0) = \partial_y v(x_0, y_0) \quad \text{and} \quad \partial_y u(x_0, y_0) = -\partial_x v(x_0, y_0).$$

PROBLEM 2.6

The result of the previous problem shows that the Cauchy Riemann equations must be satisfied at a point where the derivative exists. Show that if the real and imaginary parts of f are smooth in a neighborhood of z_0 and if the Cauchy Riemann equations are satisfied at z_0 then $f'(z_0)$ exists.

SOLUTION 2.6

We begin by forming the difference quotient at z_0

$$\frac{\Delta f}{\Delta z} = \frac{f(z_0 + \Delta z) - f(z_0)}{\Delta z}$$

$$= \frac{u(x_0 + \Delta x, y_0 + \Delta y) + iv(x_0 + \Delta x, y_0 + \Delta y) - u(x_0, y_0) - iv(x_0, y_0)}{\Delta x + i\Delta y}$$

$$= \frac{u(x_0 + \Delta x, y_0 + \Delta y) - u(x_0 + \Delta x, y_0) + u(x_0 + \Delta x, y_0) - u(x_0, y_0)}{\Delta x + i\Delta y}$$

$$+ i\frac{v(x_0 + \Delta x, y_0 + \Delta y) - v(x_0, y_0 + \Delta y) + v(x_0, y_0 + \Delta y) - v(x_0, y_0)}{\Delta x + i\Delta y}$$

Now if $u(x, y)$ and $v(x, y)$ are smooth functions in a neighborhood of z_0 then Taylor's theorem implies

$$u(x_0 + \Delta x, y_0 + \Delta y) = u(x_0 + \Delta x, y_0) + \partial_y u(x_0 + \Delta x, y_0)\,\Delta y + \text{higher order terms}$$

$$u(x_0 + \Delta x, y_0) = u(x_0, y_0) + \partial_x u(x_0, y_0)\,\Delta x + \text{higher order terms}$$

$$v(x_0 + \Delta x, y_0 + \Delta y) = v(x_0, y_0 + \Delta y) + \partial_x v(x_0, y_0 + \Delta y)\,\Delta x + \text{higher order terms}$$

$$v(x_0, y_0 + \Delta y) = v(x_0, y_0) + \partial_y v(x_0, y_0)\,\Delta y + \text{higher order terms}$$

where higher order terms are expressions involving Δx and Δy raised to powers higher than one. Then

$$\frac{\Delta f}{\Delta z} = \frac{(\partial_x u + i\partial_x v)\,\Delta x}{\Delta x + i\Delta y} + \frac{(\partial_y u + i\partial_y v)\,\Delta y}{\Delta x + i\Delta y} + \text{higher order terms}$$

$$= (\partial_x u + i\partial_x v) + \frac{1 + \dfrac{\Delta y}{\Delta x}\dfrac{\partial_y u + i\partial_y v}{\partial_x u + i\partial_x v}}{1 + \dfrac{\Delta y}{\Delta x}i} + \text{higher order terms}$$

It is clear from this expression that the difference quotient tends to the limit $\partial_x u(x_0, y_0) + i\partial_x v(x_0, y_0)$ independent of the path along which z_0 is approached if

$$\frac{\partial_y u + i\partial_y v}{\partial_x u + i\partial_x v} = i \quad \text{at } (x_0, y_0) \tag{1}$$

But this is equivalent to the Cauchy Riemann conditions. Thus $f'(z_0)$ exists and equals $\partial_x u(x_0, y_0) + i\partial_x v(x_0, y_0)$ if $u(x, y)$ and $v(x, y)$ are smooth in a neighborhood of z_0 and if the Cauchy Riemann conditions are satisfied at z_0.

PROBLEM 2.7

Show that $f'(z_0)$ may fail to exist even though the Cauchy Riemann equations are satisfied at z_0.

SOLUTION 2.7

Consider the function $f(z)$ such that $f(0) = 0$ and for z different from zero

$$f(z) = \frac{x^3 - y^3}{x^2 + y^2} + i\frac{x^3 + y^3}{x^2 + y^2} = u(x, y) + iv(x, y)$$

We compute the derivatives of u and v in order to show

$$\partial_x u(x,0) = 1 \quad \text{and} \quad \partial_x v(x,0) = 1 \quad \text{for } x \text{ not equal to zero}$$

$$\partial_y u(0,y) = -1 \quad \text{and} \quad \partial_y v(0,y) = -1 \quad \text{for y not equal to zero}$$

Then the Cauchy Riemann conditions are satisfied at $z = 0$. However,

$$\frac{\Delta f}{\Delta z} = \frac{\dfrac{(\Delta x)^3 - (\Delta y)^3}{(\Delta x)^2 + (\Delta y)^2} + i\dfrac{(\Delta x)^3 + (\Delta y)^3}{(\Delta x)^2 + (\Delta y)^2}}{\Delta x + i\Delta y}$$

$$= \frac{\dfrac{1 - (\Delta y/\Delta x)^3}{1 + (\Delta y/\Delta x)^2} + i\dfrac{1 + (\Delta y/\Delta x)^3}{1 + (\Delta y/\Delta x)^2}}{1 + i\Delta y/\Delta x}$$

and the limiting value of the difference quotient clearly depends on the path by which Δz tends to zero. For example along the line $y = kx$ the difference quotient tends to the value

$$\frac{\dfrac{1-k^3}{1+k^2} + i\dfrac{1+k^3}{1+k^2}}{1+ik} = \frac{(1-k^3) + k(1+k^3) + i((1+k^3) - k(1-k^3))}{(1+k^2)(1-k^2)}$$

which clearly depends on the value of the parameter k. Since the limiting value depends on the path, the limit fails to exist. The difficulty here lies in the fact that while the derivatives of u and v exist at the origin, they are not continuous there.

Derivatives

PROBLEM 2.8

Find $f'(z)$ at every point where it exists for each of the following functions:

(a) $f(z) = xy - ix$

(b) $f(z) = z^2 - 3z$

(c) $f(z) = 1/(z-1)$

(d) $f(z) = z - z^*$

SOLUTION 2.8

For $f(z) = xy - ix$ we compute $\partial_x u = y$, $\partial_y u = x$, $\partial_x v = -1$ and $\partial_y v = 0$. Then the derivatives are smooth and the Cauchy Riemann equations are satisfied at $x = 1, y = 0$. Then $f'(1) = -i$. Note that $f(z)$ is not analytic at (1, 0) since the Cauchy Riemann equations are satisfied only at the point and not in any neighborhood of the point.

For $f(z) = z^2 - 3z = x^2 - y^2 - 3x + i(2xy - 3y)$ we compute $\partial_x u = 2x - 3 = \partial_y v$ and $\partial_y u = -2y = -\partial_x v$. Then the derivatives are everywhere continuous and the Cauchy Riemann equations are satisfied at all points as well. Then $f'(z) = 2z - 3$ for all z and $f(z)$ in an entire function.

For $f(z) = 1/(z-1)$ we note that $f(z)$ is not defined at $z = 1$. However, for z not equal to 1 we have

$$f(z) = \frac{x-1}{(x-1)^2 + y^2} - i\frac{y}{(x-1)^2 + y^2}$$

hence

$$\partial_x u = \frac{y^2 - (x-1)^2}{((x-1)^2 + y^2)^2} = \partial_y v$$

$$\partial_y u = \frac{2y(x-1)}{((x-1)^2 + y^2)^2} = -\partial_x v$$

and

$$f'(z) = -1/(z-1)^2 \quad \text{for } z \text{ not equal to } 1.$$

Evidently $f(z) = 1/(z-1)$ is analytic everywhere except at $z = 1$. We refer to $z = 1$ as a *singular point* for the function $f(z)$.

For $f(z) = z - z^*$ we have $\partial_x u = \partial_y u = \partial_x v = 0$ and $\partial_y v = 2$. Thus the Cauchy Riemann equations are not satisfied at any point and $f'(z)$ exists nowhere.

Harmonic Conjugate

PROBLEM 2.9

Find $v(x, y)$ such that $f(z) = xy + iv(x, y)$ is analytic everywhere.

SOLUTION 2.9

In order that $f(z)$ be analytic everywhere it is necessary that the Cauchy Riemann equations be satisfied at every point. Since $\partial_x u = y$ and $\partial_y u = x$ this means that v must satisfy $\partial_x v = -x$ and $\partial_y v = y$. The first of these equations is satisfied if $v(x, y) = \varphi(y) - x^2/2$ for an arbitrary smooth function $\varphi = \varphi(y)$. Then the second condition on $v(x, y)$ implies $\varphi'(y) = y$, which is to say $\varphi(y) = y^2/2 + C$. Then for any complex constant C, the function

$$f(z) = xy + i(y^2 - x^2)/2 + iC$$

is analytic everywhere. The function $v(x, y) = (y^2 - x^2)/2$ is the harmonic conjugate of the function $u(x, y) = xy$. Note that this conjugate is unique only up to an additive constant C.

Orthogonal Families

PROBLEM 2.10

Let $f(z) = u(x, y) + iv(x, y)$ be analytic throughout a region D in the complex plane. For arbitrary constants a and b let C_a and S_b denote curves in the plane along which $u = a$ and $v = b$, respectively. Then show that at each point where curves C_a and S_b intersect, they are orthogonal to one another.

SOLUTION 2.10

Since C_a and S_b are level curves of u and v, respectively, then at each point (x, y) the normal vectors to these curves are given by

$$N_a = \mathbf{grad}\, u = (\partial_x u, \partial_y u) \quad \text{and} \quad N_b = \mathbf{grad}\, v = (\partial_x v, \partial_y v)$$

At a point (x_0, y_0) where a curve C_a crosses a curve S_b the dot product of the normals equals

$$N_a \cdot N_b = (\partial_x u, \partial_y u) \cdot (\partial_x v, \partial_y v) = \partial_x u \partial_x v + \partial_y u \partial_y v$$

But u and v are the real and imaginary parts of an analytic function, hence they are harmonic conjugates. Then the Cauchy Riemann equations imply that

$$N_a \cdot N_b = \partial_x u \partial_x v + \partial_y u \partial_y v = \partial_y v \partial_x v - \partial_x v \partial_y v = 0$$

and it follows that the normals to C_a and S_b are orthogonal; i.e., the curves intersect at right angles at the point (x_0, y_0). For example the level curves of the harmonic conjugates $u(x, y) = xy$ and $v(x, y) = (y^2 - x^2)/2$ are two families of mutually orthogonal hyperbolas in the xy-plane. (See Figure 6.4(a))

Elementary Functions

PROBLEM 2.11

Show that $f(z) = e^z = e^x(\cos y + i \sin y)$ and that Euler's formula follows when $z = iy$.

SOLUTION 2.11

If we write $f(z) = u(x,y) + iv(x,y)$ then analyticity implies that $f'(z)$ exists and equals $f'(z) = \partial_x u(x,y) + i\partial_x v(x,y)$. But then $f'(z) = f(z)$ leads to

$$\partial_x u(x,y) = u(x,y) \quad \text{and} \quad \partial_x v(x,y) = v(x,y)$$

i.e.,

$$u(x,y) = p(y)\, e^x \quad \text{and} \quad v(x,y) = q(y)\, e^x$$

for arbitrary smooth functions of y, $p(y)$ and $q(y)$. But analyticity also requires the Cauchy Riemann equations to be satisfied. Thus

$$p(y)\, e^x = q'(y)\, e^x \quad \text{and} \quad p'(y)\, e^x = -q(y)\, e^x;$$

i.e.,

$$p''(y) = -p(y) \quad \text{and} \quad q''(y) = -q(y).$$

Then $p(y) = A\cos y + B\sin y$ and $q(y) = A\sin y - B\cos y$ for A and B (at this point) arbitrary constants. Thus

$$e^z = (A\cos y + B\sin y)\, e^x + i(A\sin y - B\cos y)\, e^x$$

and for $z = x$ this reduces to $e^x = (A - iB)\, e^x$. We conclude that $A = 1$ and $B = 0$ leaving us with

$$e^z = e^x(\cos y + i\sin y)$$

Thus Euler's formula follows from the definition of the exponential function.

PROBLEM 2.12

Find all the complex numbers z such that $e^z = -2$.

SOLUTION 2.12

Note that this equation has no solution in the set of real numbers. However, for z complex we have

$$e^z = e^x(\cos y + i\sin y) = -2 = 2(\cos(\pi + 2k\pi) + i\sin(\pi + 2k\pi));$$

i.e.,

$$x = \ln 2 \quad \text{and} \quad y = (2k+1)\pi \quad \text{for any integer } k.$$

PROBLEM 2.13

Find all the points where e^{z^*} is analytic.

SOLUTION 2.13

We have

$$e^{z^*} = e^x e^{-iy} = e^x \cos y - ie^x \sin y$$

and

$$\partial_x u = e^x \cos y \qquad \partial_y u = -e^x \sin y$$

$$\partial_x v = -e^x \sin y \qquad \partial_y v = -e^x \cos y.$$

Since e^x never vanishes and $\sin y$ and $\cos y$ are never simultaneously zero, the Cauchy Riemann equations are not satisfied at any point and this function is nowhere analytic.

PROBLEM 2.14

Show that the definitions (2.7) for the complex sine and cosine are equivalent to

$$\sin z = \sin x \cosh y + i \cos x \sinh y \qquad (1)$$

$$\cos z = \cos x \cosh y - i \sin x \sinh y \qquad (2)$$

SOLUTION 2.14

The definition (2.7) for the complex sine implies

$$\begin{aligned} \sin(x+iy) &= (e^{ix-y} - e^{-ix+y})/2i \\ &= ((\cos x + i\sin x)e^{-y} - (\cos x - i\sin x)e^{y})/2i \\ &= \cos x(e^{-y} - e^{y})/2i + \sin x(e^{-y} + e^{y})/2 \\ &= i\cos x \sinh y + \sin x \cosh y. \end{aligned}$$

This is (1). Note that if we formally apply the identity for the sine of a sum, we find

$$\sin(x+iy) = \sin x \cos(iy) + \sin(iy)\cos x \qquad (3)$$

Equating real and imaginary parts in (1) and (3), we obtain

$$\cos iy = \cosh y \quad \text{and} \quad \sin iy = i\sinh y. \qquad (4)$$

In the same way we can derive (2) from the definition (2.7) for the complex cosine function. Note that the complex sine and cosine are entire functions.

PROBLEM 2.15

Show that

$$|\sin z|^2 = \sin^2 x + \sinh^2 y \quad \text{and} \quad |\cos z|^2 = \cos^2 x + \sinh^2 y \tag{1}$$

hence

$$\sin z = 0 \text{ if and only if } z = k\pi \tag{2}$$
$$\cos z = 0 \text{ if and only if } z = (2k+1)\pi/2$$

for k = integer.

SOLUTION 2.15

It follows from the definition (2.8) that

$$|\sin z|^2 = \sin^2 x \cosh^2 y + \cos^2 x \sinh^2 y$$
$$= \sin^2 x \cosh^2 y + (1 - \sin^2 x) \sinh^2 y$$
$$= \sin^2 x (\cosh^2 y - \sinh^2 y) + \sinh^2 y = \sin^2 x + \sinh^2 y$$

The other half of (1) follows similarly. Note that $\sin z = 0$ if and only if $\sin x = \sinh y = 0$. These equations are satisfied if and only if $y = 0$ and $x = k\pi$ for k any integer. Similarly $\cos z = 0$ if and only if $\cos x = \sinh y = 0$. These equations are satisfied if and only if $y = 0$ and $x = (2k+1)\pi/2$ for k any integer.

PROBLEM 2.16

Find all the complex numbers z such that $\sin 2z = 5$.

SOLUTION 2.16

Note that this equation has no real solutions since the absolute value of the real sine function never exceeds 1. However,

$$\sin 2z = \sin 2x \cosh 2y + i \sinh 2y \cos 2x = 5$$

implies

$$\sin 2x \cosh 2y = 5 \quad \text{and} \quad \sinh 2y \cos 2x = 0.$$

The second equation is satisfied if $y = 0$ or if $x = (2k+1)\pi/4$ for k an integer. The choice $y = 0$ leads to $\sin 2x = 5$ in the first equation. Since this has no solution we conclude that $x = (2k+1)\pi/4$ for k an integer. Substituting this into the first equation leads to

$$\cosh 2y = (-1)^k 5$$

which has no solution if k is odd. Then $x = (4k+1)\pi/4$ for k an integer and

$$\cosh 2y = (e^{2y} + e^{-2y})/2 = 5$$

or

$$w^2 - 10w + 1 = 0 \quad \text{where} \quad w = e^{2y};$$

i.e.,

$$e^{2y} = 5 \pm 2\sqrt{6} \quad \text{and} \quad y = \frac{1}{2}\ln|5 \pm 2\sqrt{6}|$$

Alternatively we can solve for z from the equation

$$\sin 2z = \frac{e^{i2z} - e^{-i2z}}{2i} = 5;$$

i.e.,

$$w - 1/w = 10i \quad \text{where} \quad w = e^{i2z}.$$

Then

$$w^2 - 10iw - 1 = 0 \quad \text{and} \quad z = \frac{1}{2i}\log w$$

Solving the quadratic equation for w leads to the solution for z which is identical to the solution found by the previous method.

PROBLEM 2.17 PARTIAL FRACTIONS DECOMPOSITION

Show that if $Q_n(z) = q_n(z - z_1) \ldots (z - z_n)$ has n distinct zeroes and P_m is a polynomial of degree $= m < n$ then

$$R(z) = \frac{P_m(z)}{Q_n(z)} = \frac{A_1}{z - z_1} + \ldots + \frac{A_n}{z - z_n} \tag{1}$$

where $A_k = P_m(z_k)/Q'(z_k)$, for $k = 1$ to n.

SOLUTION 2.17

If we suppose $R(z)$ has a partial fractions expansion of the form (1) for some choice of constants A_1 to A_n, then for any one of the zeroes, z_k, of $Q(z)$ we can write

$$(z - z_k)\frac{P_m(z)}{Q_n(z)} = (z - z_k)\left(\frac{A_1}{z - z_1} + \ldots + \frac{A_{k-1}}{z - z_{k-1}} + \ldots \right.$$
$$\left. + \frac{A_{k+1}}{z - z_{k+1}} + \ldots + \frac{A_n}{z - z_n}\right) + A_k \tag{2}$$

As z tends to z_k, the right hand side of (2) tends to the limit A_k. On the other hand since P_m and Q_n are entire functions

$$(z - z_k)\frac{P_m(z)}{Q_n(z)} = \frac{P_m(z)}{Q_n(z)/(z - z_k)}$$

$$= \frac{P_m(z)}{\dfrac{Q_n(z) - Q_n(z_k)}{z - z_k}} \rightarrow \frac{P_m(z_k)}{Q_n'(z_k)} \quad \text{as } z \rightarrow z_k$$

Thus

$$A_k = \frac{P_m(z_k)}{Q_n'(z_k)} \qquad \text{for } k = 1 \text{ to } n.$$

For example

$$\frac{z}{z^2+1} = \frac{P(i)}{Q'(i)}\frac{1}{z-i} + \frac{P(-i)}{Q'(-i)}\frac{1}{z+i} = \frac{1}{2}\frac{1}{z-i} + \frac{1}{2}\frac{1}{z+i}$$

PROBLEM 2.18 BRANCHES OF MULTIPLE VALUED FUNCTIONS

Let $F_1(z)$, $F_2(z)$ denote the following distinct branches of the function $\log z$

$$F_1(z) = \ln r + i\vartheta \quad \text{for} \quad z = re^{i\vartheta} \text{ in } D_1 = \{r > 0, 0 \le \vartheta < 2\pi\}$$

$$F_2(z) = \ln r + i\vartheta \quad \text{for} \quad z = re^{i\vartheta} \text{ in } D_2 = \{r > 0, -\pi \le \vartheta < \pi\}.$$

Show that for $z_0 = (-\sqrt{3} + i)/2$ we have $F_1(z_0{}^2) = 2F_1(z_0)$ but $F_2(z_0{}^2)$ does not equal $2F_2(z_0)$.

SOLUTION 2.18

Note first that

$$z_0 = 1e^{i5\pi/6} \quad \text{and} \quad z_0{}^2 = 1e^{i5\pi/3} = 1e^{-i\pi/3}$$

Then $\qquad F_1(z_0) = \ln 1 + i5\pi/6 = i5\pi/6$

and $\qquad F_2(z_0) = 0 + i5\pi/6.$

But $\qquad F_1(z_0{}^2) = i5\pi/3 = 2F_1(z_0)$

while $\qquad F_2(z_0{}^2) = -i\pi/3 \ne 2F_2(z_0).$

As this example illustrates, it is not necessarily the case that $\log(z^n)$ equals $n \log z$ for z complex although the equality holds for the real logarithm function. The difficulty in this example is due to the fact that 2 times $\arg z_0$ exceeds the upper limit of π for the argument of z in D_2. Thus in D_2 we have $\arg(z_0{}^2) = 5\pi/3 - 2\pi = -\pi/3$ which is not equal to twice $\arg z_0$.

PROBLEM 2.19

For arbitrary complex number c define $z^c = e^{c \log z}$. Then compute the principal value for each of the following:

(a) i^{2-3i}

(b) $(1-i)^{i}$

(c) $(-1+i\sqrt{3})^{1+i}$

SOLUTION 2.19

First we write

$$i = e^{i\pi/2} \quad \text{and} \quad \log i = \ln 1 + i\pi/2 = i\pi/2$$

$$i-1 = \sqrt{2}e^{i\pi/4} \quad \text{and} \quad \log(i-1) = \frac{1}{2}\ln 2 + i\pi/4$$

$$-1+i\sqrt{3} = 2e^{i2\pi/3} \quad \text{and} \quad \log(-1+i\sqrt{3}) = \ln 2 + i2\pi/3$$

Then

$$i^{2-3i} = e^{3\pi/2}e^{i\pi} = -e^{3\pi/2}$$

$$(i-1)^{i} = e^{i(\ln 2)/2}e^{-\pi/4}$$

$$(-1+i\sqrt{3})^{1+i} = e^{\ln 2 - 2\pi/3}e^{i(\ln 2 + 2\pi/3)}$$

PROBLEM 2.20

If we define: $w = \arcsin z$ if and only if $\sin w = z$
$w = \arccos z$ if and only if $\cos w = z$

then show that

$$\arcsin z = -i\log(iz + \sqrt{1-z^2}) \tag{1}$$

$$\arccos z = -i\log(z + i\sqrt{1-z^2}) \tag{2}$$

SOLUTION 2.20

Since $w = \arcsin z$ if and only if

$$z = \sin w = \frac{e^{iw} - e^{-iw}}{2i}$$

we have $e^{2iw} - 2ize^{iw} - 1 = 0$. Solving this quadratic equation for e^{iw} leads to

$$e^{iw} = iz + \sqrt{1-z^2}$$

or

$$w = -i\log(iz + \sqrt{1-z^2}).$$

We can obtain (2) similarly. We can also define $w = \arctan z$ if and only if

$$z = \frac{\sin w}{\cos w} = -i\frac{e^{iw} - e^{-iw}}{e^{iw} + e^{-iw}}$$

and this leads to

$$w = \arctan z = \frac{i}{2}\log\frac{i+z}{i-z}$$

PROBLEM 2.21

Show that
(a) $d/dz(\sin z) = \cos z$

(b) $d/dz(\log z) = 1/z$

Note that these differentiation formulas are identical to the formulas for the real valued counterparts for these two analytic functions.

SOLUTION 2.21

Write $f(z) = \sin z = \sin x \cosh y + i\cos x \sinh y$ and recall that $f'(z) = u_x + iv_x$.
Then

$$f'(z) = \cos x \cosh y - i\sin x \sinh y = \cos z.$$

To show (b) write $f(z) = \log z = \ln (x^2+y^2)^{1/2} + i\arctan(y/x)$.
Then

$$f'(z) = \frac{1}{2}\partial_x \log(x^2+y^2) + i\partial_x \arctan(y/x)$$

$$= \frac{1}{2}\frac{2x}{x^2+y^2} + i\frac{-y/x^2}{1+(y/x)^2} = \frac{x-iy}{x^2+y^2} = 1/z$$

PROBLEM 2.22

Show that

(a) $d/dz(\arcsin z) = \dfrac{1}{\sqrt{1-z^2}}$

(b) $d/dz(\operatorname{arcsinh} z) = \dfrac{1}{\sqrt{z^2+1}}$

SOLUTION 2.22

By definition, $w = \arcsin z$ if and only if $z = \sin w$. Differentiating both sides of this last equality with respect to z gives

$$1 = \cos w\frac{dw}{dz}$$

Then since $\cos^2 w + \sin^2 w = 1$ for all complex w, it follows that

$$\frac{dw}{dz} = \frac{1}{\sqrt{1 - \sin^2 w}} = \frac{1}{\sqrt{1 - z^2}}$$

Similarly, $w = \text{arc sinh} z$ if and only if $z = \sinh w$, hence

$$1 = \cosh w \frac{dw}{dz}.$$

Then the identity $\cosh^2 w - \sinh^2 w = 1$ leads to the result (b).

SUMMARY

The function $f(z) = u(x, y) + iv(x, y)$ tends to the complex limit $L = L_1 + iL_2$ as z tends to $z_0 = x_0 + iy_0$ if and only if the real valued functions u and v tend to real limits L_1 and L_2, respectively, as (x, y) tends to (x_0, y_0). Thus the limit problem for a complex valued function of a complex variable is equivalent to two limit problems for real valued functions of two real variables. Similarly the continuity of f at z_0 is equivalent to the continuity of u and v at (x_o, y_0). The differentiability of f(z) is a more delicate matter.

The function f(z) is differentiable at z_0 if the difference quotient limit (2.1) exists but existence of the limit requires a certain linkage between the various partial derivatives of the real and imaginary parts of f(z). If f(z) is differentiable at z_0 then all of the first partials u_x, u_y, v_x, v_y must exist at (x_0, y_0) and they must satisfy the Cauchy Riemann equations at that point. On the other hand, if u_x, u_y, v_x, v_y are all continuous at (x_0, y_0) and if the Cauchy Riemann equations are satisfied there then this is sufficient to imply that f(z) is differentiable at the point z_0. Then the derivative is given by any of the following equivalent expressions

$$\begin{aligned} f'(z_0) &= u_x + iv_x = v_y - iu_y \\ &= \cos\vartheta u_r - r^{-1}\sin\vartheta u_\vartheta + i(\cos\vartheta v_r - r^{-1}\sin\vartheta v_\vartheta) \\ &= \sin\vartheta v_r + r^{-1}\cos\vartheta v_\vartheta - i(\sin\vartheta u_r + r^{-1}\cos\vartheta u_\vartheta) \end{aligned}$$

If f is differentiable not just at z_0 but in an entire neighborhood of z_0, then f is analytic at z_0. If f is analytic at every point in a domain D, then f *is said to be analytic on D. Analyticity has many unexpected consequences.*

If $f(z)$ is analytic on domain D then $u = \text{Re} f$ and $v = \text{Im} f$ are harmonic on D; i.e. u and v are smooth and satisfy Laplace's equation at every point of D. Conversely, if u is harmonic on D and if v is the harmonic conjugate of u then $f = u + iv$ is analytic on D.

Polynomials in z are analytic everywhere (entire) and rational functions of z are analytic at each point where the denominator is different from zero. The functions e^z, $\sin z$, $\cos z$, $\sinh z$ and $\cosh z$ are all entire functions and the remaining trigonometric and hyperbolic functions are analytic at each point where they are defined. The functions $\log z$ and $z^{1/n}$ are multiple valued but single valued branches can be defined by placing branch cuts in the complex plane. The single valued branches are then analytic at all points in the plane with the exception of points on the branch cut.

The rules of differentiation for analytic functions of a complex variable are identical to the rules of differentiation for a real valued function of one real variable. Additionally, the differentiation formulas for the elementary analytic functions are identical to the formulas for their real valued counterparts.

SUPPLEMENTARY PROBLEMS

1. Use the definition of the derivative to show that the following functions are nowhere differentiable:
   (a) $f(z) = 1/z^*$
   (b) $f(z) = \text{Re}\, z$
   (c) $f(z) = z - z^*$
2. Use the definition of the derivative to show that
   (a) $f(z) = 6z + 2$ is differentiable for all z
   (b) $f(z) = 3/z$ is differentiable for z different from zero
3. At which points in the complex plane are the following differentiable? At what points are they analytic?
   (a) $f(z) = iz^{-2}$
   (b) $f(z) = z \text{Im}\, z$
   (c) $f(z) = y^2 - ix^2$
   (d) $f(z) = x + iy^2$
   (e) $f(z) = x^2 - iy$

(f) $f(z) = 2x^2 + i3y^3$

(g) $f(z) = r^2 e^{i2\vartheta}$

(h) $f(z) = r^2 e^{i\vartheta}$

(i) $f(z) = z^i$

4. Show in two ways that $u = u(x, y)$ is harmonic. Find the harmonic conjugate and an analytic function $f(z)$ for which $u = \text{Re} f$.

(a) $u(x, y) = 2x - x^3 + 3xy^2$

(b) $u(x, y) = 2y/(x^2 + y^2)$

5. Find all complex numbers z which satisfy

(a) $e^z = -3$

(b) $\log z = 1 + i2\pi$

(c) $\log z = i(4n + 1)\pi/2$

SOLUTIONS TO SUPPLEMENTARY PROBLEMS

3. (a) f is differentiable and analytic with $f'(z) = -2iz^{-3}$ at all z different from zero

(b) f is differentiable only at $z = 0$; f is nowhere analytic

(c) f is differentiable along the line $y = x$; f is nowhere analytic

(d) f is differentiable on the line $y = 1/2$; f is nowhere analytic

(e) f is differentiable on the line $x = -1/2$; f is nowhere analytic

(f) f is differentiable on the curve $4x = 9y^2$; f is nowhere analytic

(g) $f'(z) = 2z$ at all z; f is entire

(h) f is differentiable only at $z = 0$; f is nowhere analytic

(i) $f'(z) = iz^{i-1}$ and f is analytic at all z not on the branch cut of $\log z$

4. (a) $u(x, y) = 2x - x^3 + 3xy^2 = \text{Re}(2z - z^3)$ hence $v(x, y) = \text{Im}(2z - z^3)$

(b) $u(x, y) = 2y/(x^2 + y^2) = \text{Re}(2i/z)$ hence $v(x, y) = \text{Im}(2i/z)$

5. (a) $z_n = \ln 3 + i(2n + 1)\pi$, n = integer

(b) $z = e$

(c) $z = i$

3

Integration

Integration of a complex valued function f(z) with respect to the complex variable z involves integrating f along a curve C in the complex plane. Such integrals can be defined directly in terms of Riemann sums or they can be expressed in terms of planar contour integrals of real valued functions. We choose the second approach and begin by recalling a few key results about real contour integrals.

We show how complex contour integrals can be evaluated by parameterizing the contour and the basic properties of complex contour integrals are presented. The fundamental theorem of complex integration, the Cauchy Goursat theorem, is discussed and its many consequences developed. The material in this chapter provides the theoretical basis for later applications.

CONTOURS

PARAMETRIC DESCRIPTION OF A CURVE

Let C denote the set of points $\{z = x + iy\colon\ z(t) = (x(t), y(t)),\ a \le t \le b\}$. Then we say that C is a *curve* in the complex plane described *parametrically* in terms of the parameter t. As t varies from a to b, $z(t)$ traces out the points of C. The functions $x(t)$, $y(t)$ in the parametric description are not unique. In fact, for any given curve C, there are infinitely many different parameterizations.

ARCS

If $\boldsymbol{P}(a)$ and $\boldsymbol{P}(b)$ are distinct points then C is said to be an *arc* and if there are no values $t_1, t_2,\ a \le t_1 < t_2 \le b$, such that $z(t_1) = z(t_2)$ then we say that C is a *simple arc*; i.e., C does not cross over itself at any point.

SIMPLE CLOSED CURVES

If $z(a) = z(b)$ then we say that C is a *closed curve*; i.e., the initial and final points of C coincide. A closed curve for which $z(t_1) \neq z(t_2)$ for $a < t_1 < t_2 < b$, is called a *simple closed curve.*

SMOOTH CURVES

If the functions $x(t)$ and $y(t)$ belong to $C^1[a, b]$ we say that C is a *smooth curve*. We will always assume that there is no t in $[a, b]$ where the derivatives $x'(t)$, $y'(t)$ vanish simultaneously. This ensures that C has a continuously turning tangent line at each point.

PIECEWISE SMOOTH CURVES OR CONTOURS

More generally we say that C is *piecewise smooth* if there exists a partition $\{a = t_0, t_1, \ldots, t_n = b\}$ for $[a, b]$ such that x, y are C^1 on $[t_{k-1}, t_k]$ for $k = 1, \ldots, n$. Then each of the subarcs $C_k = \{z(t) = (x(t), y(t)), t_{k-1} \leq t \leq t_k\}$ is a smooth arc. We shall use the term *contour* to mean a simple, piecewise smooth curve. A contour whose initial and final points are the same will be called a *closed contour* and if these are the only points where the contour meets itself, we say it is a *simple closed contour.*

ORIENTATION ON A CONTOUR

Once a parameterization has been chosen for a contour $C = \{z(t) = (x(t), y(t))\ a \leq t \leq b\}$ then an *orientation* is induced on C. We say that C is traced in the *positive* sense if C is traversed from $z(a)$ to $z(b)$ and is traced in the *negative* sense from $z(b)$ to $z(a)$. We refer to $z(a)$ and $z(b)$ respectively as the *initial* endpoint and *final* endpoint for C.

ORIENTATION ON CLOSED CONTOURS

If C is a simple closed contour with $z(a) = z(b)$ then C may be traversed in either the clockwise or the counter clockwise sense. It can be shown that C separates the complex plane into two domains, one of which is bounded and is called the *interior* of C. The other domain is called the *exterior* of C and is unbounded. It is customary to say that a simple closed contour C is traversed in the positive sense if the interior of C is always on the left as the contour is traced out.

EXAMPLE 3.1

(a) Consider the contour C shown in Figure 3.1(a). One parameterization for this contour is given by:

$$x(t) = t \quad -1 \leq t \leq 1 \qquad y(t) = \begin{cases} -t & -1 \leq t \leq 0 \\ t & 0 \leq t \leq 1 \end{cases}$$

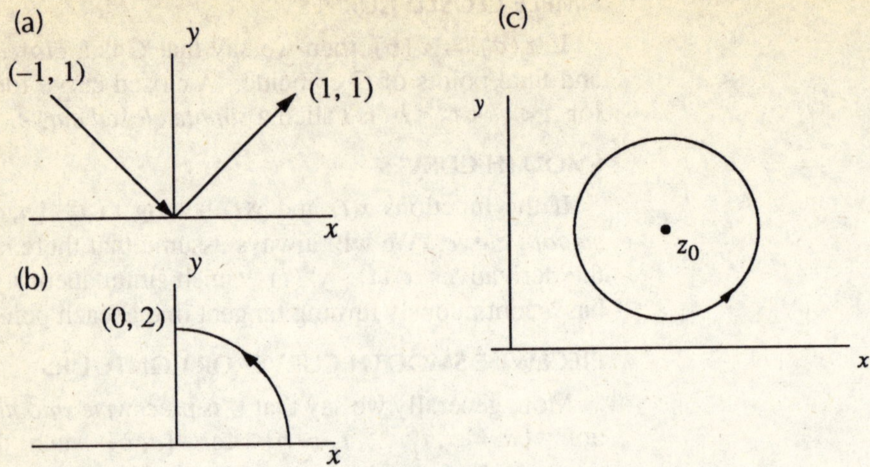

Figure 3.1

Then as t varies from -1 to 0, $(x(t), y(t))$ moves along the line $y = -x$ from $(-1, 1)$ to $(0, 0)$ and as t varies from 0 to 1, (x, y) moves along $y = x$ from $(0, 0)$ to $(1, 1)$. The initial endpoint is $(-1, 1)$ and $(1, 1)$ is the final endpoint.

(b) The simple closed contour C shown in Figure 3.1(b) may be parameterized by

$$x(t) = \begin{cases} -t & -2 \le t \le 0 \\ 0 & 0 \le t \le 2 \\ t-2 & 2 \le t \le 4 \end{cases} \qquad y(t) = \begin{cases} \sqrt{4-t^2} & -2 \le t \le 0 \\ 2-t & 0 \le t \le 2 \\ 0 & 2 \le t \le 4 \end{cases}$$

As t varies from -2 to 4, the point $(x(t), y(t))$ moves from the initial endpoint (2, 0) along the circular arc to the point (0, 2) then down the imaginary axis to the origin and finally along the real axis to the final endpoint which is again the point (2, 0). An alternative parameterization for this same curve is given by

$$x(t) = \begin{cases} -2\sin t & -\pi/2 \le t \le 0 \\ 0 & 0 \le t \le 2 \\ t-2 & 2 \le t \le 4 \end{cases} \quad y(t) = \begin{cases} 2\cos t & -\pi/2 \le t \le 0 \\ 2-t & 0 \le t \le 2 \\ 0 & 2 \le t \le 4 \end{cases}$$

(c) The circular contour shown in Figure 3.1(c) can be parameterized by

$$z(t) = z_0 + Re^{it} \quad 0 \le t \le 2\pi.$$

This is equivalent to $x(t) = x_0 + R\cos t$, $y(t) = y_0 + R\sin t$, $0 \le t \le 2\pi$. Then (x, y) starts and ends at the point $(x_0 + R, y_0)$ and moves around the circumference of the circle in the counterclockwise direction. Since the interior of the circle is then on the left, this is the positive sense.

CONTOUR INTEGRALS

LINE INTEGRALS

An integral of a real valued function along a contour in the plane is a generalization of an integral over a closed bounded interval of the real axis. Such integrals are called *line integrals* and may be reduced to standard integrals over intervals on the real axis by means of a parametric description of the contour C.

Theorem 3.1

Let $C = \{(x, y)\colon\ x = x(t),\ y = y(t),\ a \le t \le b\}$ be a piecewise smooth curve and let $F = F(x, y)$ denote a real valued function that is defined and continuous at each point of C. Then each of the following line integrals exists and is equal to the indicated integral over $a \le t \le b$

$$\int_C F dx = \int_a^b F(x(t), y(t))\, x'(t)\, dt$$

$$\int_C F dy = \int_a^b F(x(t), y(t))\, y'(t)\, dt$$

$$\int_C F ds = \int_a^b F(x(t), y(t)) \sqrt{x'(t)^2 + y'(t)^2}\, dt$$

GREEN'S THEOREM

Green's theorem is an integral identity applying to line integrals in the plane.

Theorem 3.2

Let C denote a simple closed contour with interior domain D and suppose the functions $P = P(x, y)$ and $Q = Q(x, y)$ are defined and continuously differentiable in a neighborhood of D. Then

$$\int_C P(x, y)\, dx + Q(x, y)\, dy = \iint_D (\partial_x Q(x, y) - \partial_y P(x, y))\, dxdy$$

where C is traversed in the positive sense.

COMPLEX CONTOUR INTEGRALS

Let the complex valued function $f(z) = u(x, y) + iv(x, y)$ be piecewise continuous along a contour C in the complex plane. Then we define the *complex contour integral* of f on C as follows

$$\int_C f(z)\,dz = \int_C u\,dx - v\,dy + i\int_C v\,dx + u\,dy \tag{3.1}$$

where the real line integrals on the right are to be understood in the sense of Theorem 3.1.

PROPERTIES OF CONTOUR INTEGRALS

Complex contour integrals have the following fundamental properties which follow immediately from similar properties for real line integrals.

Theorem 3.3 For any integrand $f = f(z)$ and contour C for which the integrals exist, we have

(a) If the final endpoint of C_1 is the initial endpoint of C_2 and $C_1 + C_2$ is the contour consisting of C_1 followed by C_2, then

$$\int_{C_1+C_2} f(z)\,dz = \int_{C_1} f(z)\,dz + \int_{C_2} f(z)\,dz \tag{3.2}$$

(b) If C has an orientation induced by a parameterization and if $-C$ denotes the contour traversed in the associated negative sense then

$$\int_{-C} f(z)\,dz = -\int_C f(z)\,dz \tag{3.3}$$

(c) If M denotes the maximum value for the bounded function $|f(z)|$ on C then

$$\left|\int_C f(z)\,dz\right| \le \int_C |f(z)|\,|dz| \le ML \tag{3.4}$$

where

$$L = \int_C |dz| = \text{length of } C$$

EXAMPLE 3.2

(a) Let the contour C be given by $C = \{z(t) = x(t) + iy(t),\ a \le t \le b\}$ and consider the integral

$$\int_C dz = \int_C dx + i\,dy = \int_a^b x'(t)\,dt + i\int_a^b y'(t)\,dt$$
$$= x(b) - x(a) + i(y(b) - y(a)) = z(b) - z(a).$$

Note that the value of the integral is determined by the initial and final endpoints of the contour, $z(a)$ and $z(b)$, and does not depend on the contour itself; i.e., we made no *explicit* use of the functions $x(t)$, $y(t)$ defining C. Similarly, the integral

$$\int_C z\,dz = \int_C x\,dx - y\,dy + i\int_C x\,dy + y\,dx$$

$$= \int_a^b (x(t)x'(t) - y(t)y'(t))\,dt + \int_a^b (x(t)y'(t) + y(t)x'(t))\,dt$$

$$= \int_a^b \frac{1}{2}\frac{d}{dt}(x^2 - y^2)\,dt + i\int_a^b \frac{d}{dt}(xy)\,dt$$

$$= \frac{1}{2}(x^2 - y^2)\Big|_a^b + ixy\Big|_a^b = \frac{1}{2}(z(b)^2 - z(a)^2)$$

depends only on the endpoints of the contour C and not on the contour itself. Here we used the facts

$$\frac{d}{dt}(x^2 - y^2) = 2(x(t)x'(t) - y(t)y'(t)),$$

$$\frac{d}{dt}(xy) = x(t)y'(t) + y(t)x'(t)$$

i.e., the integrands in the integrals are *exact differentials*.

(b) The integral

$$\int_C |dz| = \int_C \sqrt{dx^2 + dy^2} = \int_a^b \sqrt{x'(t)^2 + y'(t)^2}\,dt = L$$

produces the length of the contour C and, of course, depends on C itself and not just on the endpoints of C. Similarly the integral

$$\int_C z^*\,dz = \int_C xdx + ydy + i\int_C xdy - ydx$$

$$= \int_a^b (x(t)x'(t) + y(t)y'(t))\,dt + i\int_a^b (x(t)y'(t) - y(t)x'(t))\,dt$$

$$= \frac{1}{2}(x(t)^2 + y(t)^2)\Big|_a^b + i\int_a^b (x(t)y'(t) - y(t)x'(t))\,dt$$

depends explicitly on the path since the integrand in the imaginary part of the integral is not an exact differential. This integral can only be evaluated by parameterizing C and using Theorem 3.1. Note that z^* is not analytic.

INTEGRAL THEOREMS

We will present a number of results relating to complex contour integrals, most of which are consequences of a single theorem known as the Cauchy-Goursat theorem. First we introduce some additional terminology.

SIMPLY CONNECTED DOMAINS

A domain D in the complex plane is said to be *simply connected* if any simple closed curve C in D contains only points of D in its interior. A domain which is not simply connected is said to be *multiply connected.* The disc $|z - z_0| < R$ is a simply connected domain while the annulus $R_1 < |z - z_0| < R_2$ is multiply connected.

THE CAUCHY GOURSAT THEOREM

Theorem 3.4

Suppose $f(z)$ is analytic throughout the simply connected domain D. Then for every closed contour C lying in D we have

$$\int_C f(z)\,dz = 0$$

Often this theorem is stated with the additional hypotheses that C is a simple closed contour and $f'(z)$ is continuous in D. These hypotheses lead to a simpler proof but are not necessary for the conclusion to hold.

DEFORMATION OF CONTOURS

It is a consequence of the Cauchy Goursat theorem that the contour in certain contour integrals may be altered without affecting the value of the integral.

Theorem 3.5

Suppose $f(z)$ is analytic at each point of the contours C_1 and C_2. Then

$$\int_{C_1} f(z)\,dz = \int_{C_2} f(z)\,dz$$

if C_1 and C_2 are simple closed contours with f analytic at all points between C_1 and C_2 or if C_1 and C_2 are arcs with the same initial and final endpoints and $f(z)$ is analytic at all points between C_1 and C_2.

PATH INDEPENDENCE AND ANTIDERIVATIVES

Theorem 3.5 implies that certain contour integrals are *path independent*; i.e., the value of the contour integral depends only on the endpoints of the contour.

Theorem 3.6

Suppose $f(z)$ is analytic throughout the simply connected domain D. Then for any contour C lying in D with initial endpoint z_0 and final endpoint z

$$F(z) = \int_C f(\xi)\, d\xi = \int_{z_0}^{z} f(\xi)\, d\xi$$

is a single valued analytic function of z not depending on C satisfying $F'(z) = f(z)$ at each point z in D.

Any function $F(z)$ satisfying $F'(z) = f(z)$ is called an *antiderivative* of $f(z)$. The next result shows that the fundamental theorem of calculus extends to complex contour integrals having integrands which are analytic functions.

Theorem 3.7

Suppose $f(z)$ is analytic throughout the simply connected domain D and that $F(z)$ is an antiderivative of $f(z)$. Then for any z_0, z_1 in D

$$\int_{z_0}^{z_1} f(z)\, dz = F(z_1) - F(z_0)$$

This theorem asserts that when the integrand is analytic, a contour integral can be evaluated by the fundamental theorem of calculus without having to parameterize the path of integration. However, when the integrand is not analytic then the integral must be done as a contour integral, parameterizing the path of integration and applying Theorem 3.1.

THE CAUCHY INTEGRAL FORMULA

The Cauchy Goursat theorem implies the following result known as the *Cauchy integral formula.*

Theorem 3.8

Suppose $f(z)$ is analytic in the simply connected domain D. For C any positively oriented simple closed contour in D and for any point z_0 in the interior of C

$$f(z_0) = \frac{1}{2\pi i}\int_C \frac{f(z)}{z - z_0}\, dz \tag{3.5}$$

Corollary 3.9 For each nonnegative integer n,

$$f^{(n)}(z_0) = \frac{n!}{2\pi i}\int_C \frac{f(z)}{(z - z_0)^{n+1}}\, dz \tag{3.6}$$

Corollary 3.10 If $C = \{|z - z_0| = R\}$ and M denotes the maximum value of $|f(z)|$ for z on C then

$$\left|f^{(n)}(z_0)\right| \le \frac{Mn!}{R^n} \quad \text{for } n = 0, 1, \ldots \tag{3.7}$$

The results (3.6) and (3.7) are referred to as the *extended Cauchy formula* and the *Cauchy estimates*, respectively. The next theorem is an especially important consequence of the Cauchy integral formula.

Theorem 3.11 If $f(z)$ is analytic at a point z_0 then f has derivatives of all orders at z_0 and $f^{(n)}(z)$ is analytic at z_0 for every positive integer n.

MORERA'S THEOREM

Using Corollary 3.10 we can prove a converse to the Cauchy Goursat theorem known as *Morera's theorem.*

Theorem 3.12 Suppose $f(z)$ is continuous in the simply connected domain D and that for every simple closed contour C inside D we have

$$\int_C f(z)\,dz = 0.$$

Then $f(z)$ is analytic in D.

THE MEAN VALUE THEOREM

Every analytic function has the *mean value property.* That is, if $f(z)$ is analytic in domain D then the value of $f(z)$ at any point z_0 in D is equal to the integral average of f over the circumference of any circle with center at z_0 and contained in D. That is, if $D_R(z_0)$ is contained in D, then

$$f(z_0) = \frac{1}{2\pi}\int_0^{2\pi} f(z_0 + Re^{i\vartheta})\,d\vartheta = \frac{1}{L_C}\int_C f\,ds \tag{3.8}$$

where $C = \{z = z_0 + Re^{i\vartheta}\}$, $ds = R\,d\vartheta$ and $L_C = 2\pi R$. A similar statement can be made about any real valued function $u = u(x, y)$ that is harmonic in D.

THE MAXIMUM PRINCIPLE

The mean value theorem implies the so called *maximum principle.*

Theorem 3.13 Suppose $f(z)$ is continuous on the closed bounded set D and $f(z)$ is analytic on the domain interior to the simple closed curve C that is the boundary of D. Let M denote the maximum value assumed by the continuous function $|f(z)|$ on the closed bounded set C. Then for every z inside C we have $|f(z)| \le M$. Equality occurs for some z inside C if and only if $f(z)$ is constant on D.

THE MAXIMUM-MINIMUM PRINCIPLE FOR HARMONIC FUNCTIONS

Corollary 3.14 Suppose $u(x, y)$ is continuous on the closed bounded set D and $u(x, y)$ is harmonic on the domain interior to the simple closed curve C that is the boundary of D. Let M and m denote, respectively, the maximum and minimum values assumed by $u(x, y)$ on the closed bounded set C. Then for every (x, y) inside C we have $m \le u(x, y) \le M$. Equality occurs for some (x, y) inside C if and only if $u(x, y)$ is constant on D.

THE FUNDAMENTAL THEOREM OF ALGEBRA

Using the Cauchy estimates we can prove the following theorem known as *Liouville's theorem.*

Theorem 3.15

If $f(z)$ is an entire function and is bounded for all z in the complex plane then $f(z)$ must be constant.

This result leads immediately to the result we call the *fundamental theorem of algebra.*

Theorem 3.16

The polynomial $P_n(z) = z^n + a_{n-1}z^{n-1} + \ldots + a_1 z + a_0$ of degree n has precisely n zeroes $z_1, \ldots, z_n$ if repeated roots are counted according to their multiplicity.

SOLVED PROBLEMS

Contour Integrals

PROBLEM 3.1

Use two different parameterizations of C, the counterclockwise half circle from $-2i$ to $2i$, to compute the value of the contour integral

$$\int_C |z|\, dz$$

SOLUTION 3.1

First we may parameterize C by letting $C = \{(x, y)\colon\ x(t) = 2\sin t, y(t) = -2\cos t, 0 \le t \le \pi\}$. Then $|z| = 2$, $dz = 2(\cos t + i\sin t)\, dt$ and

$$\int_C |z|\, dz = \int_0^\pi 4(\cos t + i\sin t)\, dt = 8i$$

We can also parameterize C by letting $C = \{z\colon\ z = 2e^{i\vartheta}, -\pi/2 \le \vartheta \le \pi/2\}$. Then $dz = 2ie^{i\vartheta} d\vartheta$ and

$$\int_C |z|\, dz = \int_{-\pi/2}^{\pi/2} 4ie^{i\vartheta} d\vartheta = 8i$$

PROBLEM 3.2

Let C denote the positively oriented boundary of the wedge shown in Figure 3.2. Start at the origin and compute the value of the integral

$$\int_C (z^2 + z^*)\,dz$$

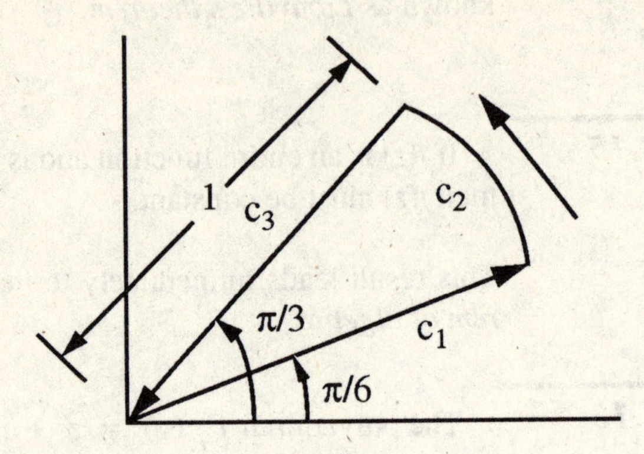

Figure 3.2

SOLUTION 3.2

The contour C is composed of three pieces; let C_1 denote the ray $\vartheta = \pi/6$ with $0 \le r \le 1$, let C_2 denote the arc $r = 1$, $\pi/6 \le \vartheta \le \pi/3$ and let C_3 denote the ray $\vartheta = \pi/3$ with r running from 1 down to 0. Then the positively oriented boundary of the wedge consists of C_1 followed by C_2 followed by C_3; i.e., $C = C_1 + C_2 + C_3$ and by (3.2) the integral over C is equal to the sum of the integrals over the three pieces

$$\int_C (z^2 + z^*)\,dz = \int_{C_1} ((re^{i\pi/6})^2 + re^{-i\pi/6})\,e^{i\pi/6}\,dr$$

$$+ \int_{C_2} ((e^{i\vartheta})^2 + e^{-i\vartheta})\,ie^{i\vartheta}\,d\vartheta + \int_{C_3} ((re^{i\pi/3})^2 + re^{-i\pi/3})\,e^{i\pi/3}\,dr$$

$$= e^{i\pi/2}\int_0^1 r^2\,dr + \int_0^1 r\,dr + i\int_{\pi/6}^{\pi/3} (e^{3i\vartheta} + 1)\,d\vartheta$$

$$+ e^{i\pi}\int_1^0 r^2\,dr + \int_1^0 r\,dr = i\pi/6$$

PROBLEM 3.3

Let C_1 denote the straight line joining (0, 0) to (1, 1) and let C_2 denote the polygonal path along the real axis from (0, 0) to (1, 0) and then along a vertical line from (1, 0) to (1, 1). Evaluate the integral $\int x\,dz$ along each contour.

SOLUTION 3.3

We can parameterize C_1 by letting $C_1 = \{(x, y): x = t, y = t, 0 \le t \le 1\}$. Then $dz = dz + idy = (1+i)\,dt$ and

$$\int_{C_1} xdz = \int_0^1 t(1+i)\,dt = (1+i)/2$$

The contour C_2 can be parameterized as follows

$$x(t) = \begin{cases} t & 0 \le t \le 1 \\ 1 & 1 \le t \le 2 \end{cases} \qquad y(t) = \begin{cases} 0 & 0 \le t \le 1 \\ t-1 & 1 \le t \le 2 \end{cases}$$

Then $dz = dt$ for $0 \le t \le 1$ and $dz = idt$ for $1 \le t \le 2$, hence

$$\int_{C_2} xdz = \int_0^1 tdt + \int_1^2 1idt = \frac{1}{2} + i$$

Note that the integral on C_1 does not equal the integral on C_2 even though the contours have the same endpoints. When the integrand is analytic then the integral is path independent. This integrand is not analytic hence path independence does not follow.

PROBLEM 3.4

For the given complex number z_0 and $R > 0$, let C denote the positively oriented circle $|z - z_0| = R$ and compute the integral

$$\int_C (z - z_0)^p dz \qquad \text{for } p \text{ an integer.}$$

SOLUTION 3.4

Here the contour C is parameterized by $C = \{z: \quad z = z_0 + Re^{i\vartheta}, 0 \le \vartheta \le 2\pi\}$. Then $dz = iRe^{i\vartheta} d\vartheta$ on C and

$$\int_C (z - z_0)^p dz = \int_0^{2\pi} (Re^{i\vartheta})^p iRe^{i\vartheta} d\vartheta = iR^{p+1} \int_0^{2\pi} e^{i(p+1)\vartheta} d\vartheta.$$

Since $e^{i2n\pi} = e^0 = 1$ for all nonzero integers n, it follows that the value of the integral is zero except when $p = -1$ in which case it equals $2\pi i$.

PROBLEM 3.5

Let C denote the positively oriented circle $|z| = R > 0$ and integrate one branch of the double valued function $f(z) = z^{-1/2}$ around C. Choose $f(z)$ to be the branch for which $f(1) = -1$.

SOLUTION 3.5

Recall that

$$f_n(z) = z^{-1/2} = r^{-1/2} e^{-i(\vartheta + 2n\pi)/2} \qquad n = 0, 1;$$

i.e.,

$$f_0(z) = r^{-1/2}e^{-i\vartheta/2}, \; f_1(z) = r^{-1/2}e^{-i(\pi+\vartheta/2)} = -r^{-1/2}e^{-i\vartheta/2}$$

The branch $f_0(z)$ corresponds to restricting the domain of $f(z)$ to $-\pi \le \vartheta < \pi$ and $f_1(z)$ corresponds to the domain $\pi \le \vartheta < 3\pi$. This is equivalent to placing a branch cut on the negative real axis. Since $f_0(1) = 1$ and $f_1(1) = -1$, the problem calls for us to integrate $f_1(z)$ around $C = \{z: z = Re^{i\vartheta}, 0 \le \vartheta \le 2\pi\}$. Then

$$\int_{C_1} f_1(z)\,dz = \int_0^{2\pi} -R^{-1/2}e^{-i\vartheta/2}(iRe^{i\vartheta})\,d\vartheta = -iR^{1/2}\int_0^{2\pi} e^{i\vartheta/2}\,d\vartheta$$

$$= -i\sqrt{R}(e^{i\pi}-1) = 4\sqrt{R}.$$

Note that the integral does not equal zero since the integrand is not analytic at every point of the contour C. The integrand fails to be analytic at the point where the contour crosses the branch cut. Note also that the integral of $f_0(z)$ around C is not equal to the integral of $f_1(z)$ around C.

PROBLEM 3.6

Show that if $f(z)$ is piecewise continuous on the contour C then

$$\left|\int_C f(z)\,dz\right| \le \int_C |f(z)|\,|dz| \le ML \tag{1}$$

where L denotes the length of C and $|f(z)| \le M$ for z on C.

SOLUTION 3.6

Let I denote the value of the integral $\int_C f(z)\,dz$. Then $I = |I|e^{i\alpha}$ for some α and

$$|I| = e^{-i\alpha}I = \int_C e^{-i\alpha}f(z)\,dz = \int_a^b e^{-i\alpha}f(z(t))\,z'(t)\,dt \tag{2}$$

where $\{z = z(t),\ a \le t \le b\}$ denotes a parameterization for C. Let $e^{-i\alpha}f(z(t))\,z'(t) = u(t) + iv(t)$ and note that

$$|u(t)| = \left|e^{-i\alpha}f(z(t))\,z'(t)\,dt\right| \le |f(z(t))|\,|z'(t)| \tag{3}$$

$$|I| = \int_a^b u(t)\,dt + i\int_a^b v(t)\,dt \tag{4}$$

Since $|I|$ is real, (4) implies

$$|I| = \int_a^b u(t)\,dt$$

Using this together with (3) leads to

$$|I| \leq \int_a^b |u(t)|\,dt \leq \int_a^b |f(z(t))||z'(t)|\,dt \leq M\int_a^b |z'(t)|\,dt = ML$$

Here we used the fact that

$$\int_a^b |z'(t)|\,dt = \int_a^b \sqrt{x'(t)^2 + y'(t)^2}\,dt = L$$

PROBLEM 3.7

Show that if $f(z)$ is analytic and $f'(z)$ is continuous throughout a simply connected domain D, then

$$\int_C f(z)\,dz = 0 \tag{1}$$

for every simple closed contour C in D.

SOLUTION 3.7

From (3.1) we have

$$\int_C f(z)\,dz = \int_C u\,dx - v\,dy + i\int_C v\,dx + u\,dy$$

The assumption that $f'(z)$ is continuous in D permits us to use Theorem 3.2, Green's theorem to write

$$\int_C f(z)\,dz = \iint_\Omega -(\partial_x v + \partial_y u)\,dxdy + i\iint_\Omega (\partial_x u - \partial_y v)\,dxdy \tag{2}$$

where Ω denotes the domain interior to C. Since $f(z)$ is analytic in D, the Cauchy Riemann equations are satisfied throughout D. This implies that both of the double integrals in (2) have integrands equal to zero and (1) follows. This proof of the Cauchy Goursat theorem requires the hypothesis that $f'(z)$ is continuous in D. We can also eliminate the hypothesis that the domain D is simply connected by assuming that the closed contour C contains only points of D in its interior.

PROBLEM 3.8

Suppose $f(z)$ is analytic throughout the simply connected domain D and let z_1, z_2 denote arbitrary points in D. Then show that for any two contours C_1 and C_2 in D joining z_1 and z_2,

$$\int_{C_1} f(z)\,dz = \int_{C_2} f(z)\,dz \tag{1}$$

SOLUTION 3.8

If C_1 and C_2 join z_1 to z_2 then $-C_2$ joins z_2 to z_1 and $C_1 - C_2$ is a closed contour in D. Since f is analytic throughout the simply connected domain D it follows by the Cauchy Goursat theorem that

$$\int_{C_1 - C_2} f(z)\,dz = 0 \tag{2}$$

Then Theorem 3.3 implies

$$\int_{C_1} f(z)\,dz - \int_{C_2} f(z)\,dz = 0 \tag{3}$$

which is clearly equivalent to (1). We have proved that the value of the integral of f over the contour C_1 is not altered if we replace C_1 by any contour C_2 with the same endpoints and such that $f(z)$ is analytic in the bounded domain inside the closed contour $C_1 - C_2$. We may also understand this to mean that the value of the integral depends only on the endpoints of the contour and not on the contour itself.

PROBLEM 3.9

Let the simple closed contour C_1 be contained inside a second simple closed contour C_2. Show that if $f(z)$ is analytic on the contours and throughout the annular domain between the two contours then

$$\int_{C_1} f(z)\,dz = \int_{C_2} f(z)\,dz \tag{1}$$

provided C_1 and C_2 have the same orientation.

SOLUTION 3.9

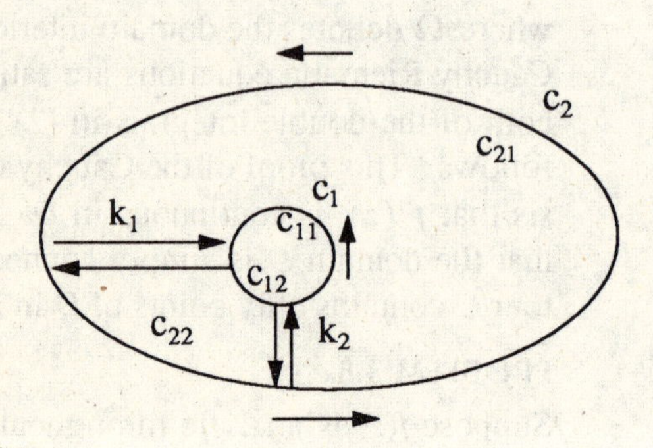

Figure 3.3

Let C_1 and C_2 be as shown in Figure 3.3. It can be shown that it is always possible construct arcs K_1, K_2 to be nonintersecting contours from C_1 to C_2. The points at which K_1, K_2 meet C_1 and C_2 induce a partition of C_1 and C_2

$$C_1 = C_{11} + C_{12} \quad \text{and} \quad C_2 = C_{21} + C_{22}.$$

We can now form two positively oriented simple closed contours

$$S_1 = C_{21} + K_1 - C_{11} - K_2 \quad \text{and} \quad S_2 = C_{22} + K_2 - C_{12} - K_1$$

for which the hypotheses imply that f is analytic on and inside each contour. The Cauchy Goursat theorem implies the integral of f around each contour is zero hence

$$\int_{S_1} f dz + \int_{S_2} f dz = \int_{S_1 + S_2} f dz = 0;$$

i.e.,

$$\int_{S_1 + S_2} f dz = \int_{C_{21} + K_1 - C_{11} - K_2 + C_{22} + K_2 - C_{12} - K_1} f dz$$

$$= \int_{C_{21} - C_{11} + C_{22} - C_{12}} f dz = \int_{C_2 - C_1} f dz = 0$$

Then

$$\int_{C_2 - C_1} f dz = \int_{C_2} f dz - \int_{C_1} f dz = 0$$

which is equivalent to (1).

PROBLEM 3.10

Suppose $f(z)$ is analytic throughtout the simply connected domain D. Show that for any contour C lying in D with initial endpoint z_0 and final endpoint z

$$F(z) = \int_C f(\xi)\, d\xi = \int_{z_0}^{z} f(\xi)\, d\xi$$

is a single valued analytic function of z not depending on C satisfying $F'(z) = f(z)$ at each point z in D.

SOLUTION 3.10

The result of Problem 3.8 shows that $F(z)$ is single valued. To show $F(z)$ is analytic with derivative equal to f we must show that for each z_1 in D

$$\lim_{z \to z_1} \frac{F(z) - F(z_1)}{z - z_1} = f(z_1);$$

i.e., we must show that for each $\varepsilon > 0$ there exists a $\delta > 0$ such that

$$\left| \frac{F(z) - F(z_1)}{z - z_1} - f(z_1) \right| < \varepsilon \text{ for all } z \text{ in } D \text{ such that } |z - z_1| < \delta \quad (1)$$

Note that

$$F(z) - F(z_1) = \int_{z_0}^{z} f(\xi)\, d\xi - \int_{z_0}^{z_1} f(\xi)\, d\xi = \int_{z_1}^{z} f(\xi)\, d\xi$$

$$= \int_{z_1}^{z} (f(\xi) - f(z_1) + f(z_1))\, d\xi$$

$$= \int_{z_1}^{z} (f(\xi) - f(z_1))\, d\xi + f(z_1) \int_{z_1}^{z} d\xi$$

$$= \int_{z_1}^{z} (f(\xi) - f(z_1))\, d\xi + f(z_1)\,(z - z_1)$$

hence

$$\frac{F(z) - F(z_1)}{z - z_1} - f(z_1) = \int_{z_1}^{z} (f(\xi) - f(z_1))\, d\xi / (z - z_1)\,. \tag{2}$$

Since $f(z)$ is analytic in D it must be continuous in D thus for each $\varepsilon > 0$ there exists a $\delta > 0$ such that

$$|f(z) - f(z_1)| < \varepsilon \quad \text{for all } z, z_1 \text{ in } D \text{ such that } |z - z_1| < \delta \tag{3}$$

Then (3), together with (3.4) implies that for $|z - z_1| < \delta$

$$\left|\int_{z_1}^{z} (f(\xi) - f(z_1))\, d\xi\right| \le \varepsilon |z - z_1| \tag{4}$$

and (4) combined with (2) gives (1).

PROBLEM 3.11

Evaluate the following integrals:

(a) $\int_1^{3i} (\sinh 3z)\, dz$

(b) $\int_C \frac{1}{z^2 + 1}\, dz$ where C denotes the positively oriented circle $|z - i| = 1$

SOLUTION 3.11

$\sinh 3z$ is an entire function and thus we are entitled to use Theorem 3.7 to evaluate the integral (a)

$$\begin{aligned}\int_1^{3i} (\sinh 3z)\, dz &= \frac{1}{3} \cosh 3z \big|_1^{3i} \\ &= \frac{1}{3} (\cosh 3x \cosh 3y + i \sinh 3x \sin 3y) \big|_{(1,0)}^{(0,3)} \\ &= \frac{1}{3} (\cos 9 - \cosh 3)\end{aligned}$$

The integrand in integral (b) is analytic at all z except $z = i, -i$ and one of these points, $z = i$, is contained in the positively oriented circle $C = \{|z - i| = 1\}$. Using the partial fractions decomposition described on Problem 2.17, we can write

$$\int_C \frac{1}{z^2 + 1}\, dz = \int_C \frac{1}{2i} \left(\frac{1}{z - i} - \frac{1}{z + i}\right) dz$$

But Problem 3.4 (with $p = -1$) and Theorem 3.4, respectively, imply

$$\int_C \frac{1}{z - i}\, dz = 2\pi i \quad \text{and} \quad \int_C \frac{1}{z + i}\, dz = 0$$

Then

$$\int_C \frac{1}{z^2+1}\,dz = \frac{1}{2i}(2\pi i - 0) = \pi$$

PROBLEM 3.12

Evaluate the integral: $$\int_C \frac{5z-i}{z^2+1}\,dz \qquad (1)$$

where C is the positively oriented circle $|z| = 2$.

SOLUTION 3.12

The integrand in this integral is analytic at all z except $z = i, -i$, both of which are inside the contour C. Since the integrand is not analytic inside C, Theorem 3.4 cannot be applied directly. We will modify the contour so as to permit us to apply the theorem. Let C_1 and C_2 denote, respectively, the left and right half of the circle $|z| = 2$; for ε a small positive constant, let $C_3 = \{|z-i| = \varepsilon\}$, $C_4 = \{|z+i| = \varepsilon\}$ and let K_1, K_2 denote the straight lines joining i to $2i$ and $-i$ to $-2i$, respectively. Then, as shown in figure 3.4, the integrand f is analytic inside the simple closed contour $C_1 + K_2 - C_4 - K_2 + C_2 - K_1 - C_3 + K_1$.

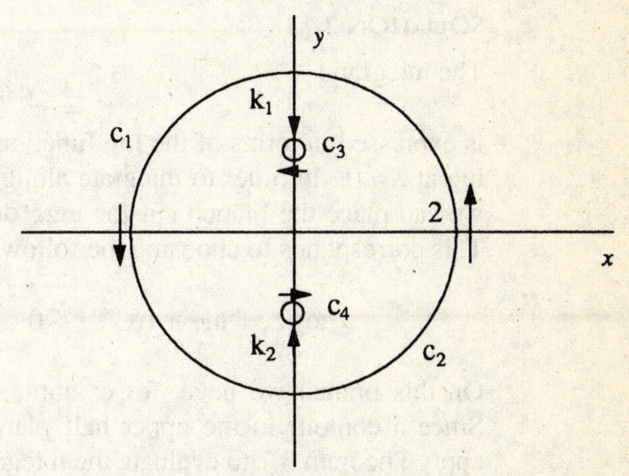

Figure 3.4

It follows from Theorem 3.4 that the integral of f around this contour is zero, and thus

$$\int_{C_1+K_2-C_4-K_2+C_2-K_1-C_3+K_1} f\,dz = \int_{C_1+C_2-C_3-C_4} f\,dz = 0;$$

i.e., $$\int_C f\,dz - \int_{C_3} f\,dz - \int_{C_4} f\,dz = 0 \qquad (2)$$

If we use partial fractions to write

$$f(z) = \frac{5z-i}{z^2+1} = \frac{2}{z-i} + \frac{3}{z+i}$$

then, using the previous problem as a guide, we see that

$$\int_{C_3} f(z)\,dz = 2\,(2\pi i) + 3\,(0) = 4\pi i$$

$$\int_{C_4} f(z)\,dz = 2\,(0) + 3\,(2\pi i) = 6\pi i$$

Then (2) implies

$$\int_{C} f(z)\,dz = 4\pi i + 6\pi i = 10\pi i.$$

The procedure of introducing "cuts" like K_1, K_2 and "circular incisions" like C_3, C_4 around singular points permits the use of the Cauchy Gourat theorem in many cases where the integrand is not analytic at every point inside the closed contour C.

PROBLEM 3.13

Evaluate the integral: $\int_{-1}^{1} z^i dz$

(a) along a contour in the upper half plane
(b) along a contour in the lower half plane.

SOLUTION 3.13

The integrand

$$z^i = e^{i\log z}$$

is expressed in terms of the log function which has a branch cut originating at $z = 0$. In order to integrate along a contour in the upper half plane we can place the branch cut for $\log z$ on the nonpositive imaginary axis. This corresponds to choosing the following branch for $\log z$

$$\log z = \ln r + i\vartheta \qquad r > 0 \qquad -\pi/2 < \vartheta \le 3\pi/2$$

On this branch we have, for example, $(-1)^{i+1} = (e^{i\pi})^{i+1} = -e^{-\pi}$. Since a contour in the upper half plane avoids this branch cut we can apply Theorem 3.7 to evaluate the integral

$$\int_{-1}^{1} z^i dz = \left.\frac{z^{i+1}}{i+1}\right|_{-1}^{1} = \frac{1}{i+1}\left((e^0)^{i+1} - (e^{i\pi})^{i+1}\right)$$

$$= \frac{1-(-e^{-\pi})}{i+1} = \frac{1}{2}(1+e^{-\pi})(1-i)$$

In order to consider a contour in the lower half plane we can place the branch cut on the nonnegative imaginary axis. This corresponds to choosing

$$\log z = \ln r + i\vartheta \qquad r > 0 \qquad -3\pi/2 < \vartheta \le \pi/2$$

Then

$$(-1)^{i+1} = (e^{-i\pi})^{i+1} = -e^{\pi}$$

and thus for a contour in the lower half plane we obtain

$$\int_{-1}^{1} z^i dz = \frac{1}{2}(1+e^{\pi})(1-i).$$

PROBLEM 3.14

Suppose $f(z)$ is analytic in the simply connected domain D. Then show that for any simple closed curve C in D and for any z_0 inside C

$$f(z_0) = \frac{1}{2\pi i}\int_C \frac{f(z)}{z-z_0}dz \tag{1}$$

SOLUTION 3.14

Since z_0 is inside C we can choose $r > 0$ sufficiently small that $C_r = \{|z-z_o| = r\}$ is contained inside C and then by Theorem 3.5

$$\int_C \frac{f(z)}{z-z_0}dz = \int_{C_r}\frac{f(z)}{z-z_0}dz \tag{2}$$

Now

$$\int_{C_r}\frac{f(z)}{z-z_0}dz = \int_{C_r}\frac{f(z)-f(z_0)}{z-z_0} + \frac{f(z_0)}{z-z_0}dz$$

$$= \int_{C_r}\frac{f(z)-f(z_0)}{z-z_0}dz + f(z_0)\int_{C_r}\frac{1}{z-z_0}dz$$

$$= \int_{C_r}\frac{f(z)-f(z_0)}{z-z_0}dz + f(z_0)\,2\pi i \tag{3}$$

In addition, since f is analytic, f is continuous hence for each $\varepsilon > 0$ there exists a $\delta > 0$ such that $|f(z) - f(z_0)| < \varepsilon$ for all z such that $|z - z_0| < \delta$. Then choosing r such that $0 < r < \delta$ we have

$$\left|\int_{C_r}\frac{f(z)-f(z_0)}{z-z_0}dz\right| \le \int_{C_r}\frac{|f(z)-f(z_0)|}{|z-z_0|}|dz| \le \frac{\varepsilon}{r}2\pi r = 2\pi\varepsilon \tag{4}$$

and using this result in (3) leads to

$$\left|\frac{1}{2\pi i}\int_{C_r}\frac{f(z)}{z-z_0}dz - f(z_0)\right| \le \varepsilon \tag{5}$$

In view of (2), (5) implies (1) since $\varepsilon > 0$ is arbitrary.

PROBLEM 3.15

Evaluate the integral $\int_C \frac{dz}{z^3(z-2)^2}$ (1)

for the positively oriented contours:

(a) $C = \{|z-2| = 1\}$

(b) $C = \{|z| = 1\}$

SOLUTION 3.15

The integrand in (1) has singularities at $z = 0$ and $z = 2$. Only the singularity at $z = 2$ is inside contour (a) hence we can apply the Cauchy integral formula, (3.6), with $n = 1$, $z_0 = 2$ and $f(z) = z^{-3}$. Then

$$\int_C \frac{dz}{z^3(z-2)^2} = 2\pi i f'(2) = -3\pi i/8 \quad C = \{|z-2| = 1\}$$

Only the singularity at $z = 0$ is inside the contour (b) hence we apply (3.6) with $n = 2$, $z_0 = 0$ and $f(z) = (z-2)^{-2}$. Then

$$\int_C \frac{dz}{z^3(z-2)^2} = \frac{2\pi i}{2!} f''(0) = 3\pi i/8 \quad C = \{|z| = 1\}$$

PROBLEM 3.16

Suppose $f(z)$ is analytic in domain D. Show that f has the mean value property in D; i.e., for any z_0 in D and any disc $D_R(z_0) = \{z: |z-z_0| \le R\}$ contained in D

$$f(z_0) = \frac{1}{2\pi}\int_0^{2\pi} f(z_0 + Re^{i\vartheta})\, d\vartheta = \frac{1}{L_C}\int_C f ds \quad (1)$$

where $L_C = 2\pi R$ is the length of the circle $|z-z_0| = R$ and $ds = R d\vartheta$.

SOLUTION 3.16

The Cauchy integral formula (3.5) implies

$$f(z_0) = \frac{1}{2\pi i}\int_C \frac{f(z)}{z-z_0} dz \quad \text{where} \quad C = \{|z-z_0| = R\}$$

On C we have $z = z_0 + Re^{i\vartheta}$ for $0 < \vartheta < 2\pi$ and $dz = iRe^{i\vartheta} d\vartheta$, hence

$$f(z_0) = \frac{1}{2\pi i}\int_0^{2\pi} \frac{f(z_0 + Re^{i\vartheta})}{Re^{i\vartheta}} iRe^{i\vartheta} d\vartheta$$

$$= \frac{1}{2\pi R}\int_0^{2\pi} f(z_0 + Re^{i\vartheta})\, R d\vartheta$$

This is (1) and it asserts that the value of f at z_0 equals the integral average of $f(z)$ over a circle with center at z_0. Note that by equating real parts on the two sides of (1) we obtain

$$u(x_0, y_0) = \frac{1}{2\pi R}\int_0^{2\pi} u(z_0 + R\cos\vartheta, y_0 + R\sin\vartheta)\, ds$$

where $ds = Rd\vartheta$ denotes the element of arclength on C. Thus harmonic functions also have the mean value property.

PROBLEM 3.17

Let C denote the boundary of the bounded domain D in the xy-plane. Suppose $u = u(x, y)$ is C^2 in D and is continuous on the union of D and C. Suppose further that u satisfies

$$\partial_{xx}u + \partial_{yy}u = 0 \text{ in } D \quad \text{and} \quad u = f \text{ on } C \tag{1}$$

for f a given function that is defined and continuous on C. Then we say u is a solution of the *Dirichlet boundary value problem* for Laplace's equation on D. Show that there is at most one solution for (1).

SOLUTION 3.17

Suppose that there are two functions, $u = u(x, y)$ and $v = v(x, y)$, each satisfying (1). Then $w = u - v$ satisfies

$$\partial_{xx}w + \partial_{yy}w = 0 \text{ in } D \quad \text{and} \quad w = 0 \text{ on } C \tag{2}$$

Then since w is harmonic in D we can apply Corollary 3.14 to conclude that $m \le w(x, y) \le M$ for all (x, y) in D and on C. But $w = 0$ on C and thus $m = M = 0$. It follows that w is identically zero in D and $u = v$ in D.

PROBLEM 3.18

Suppose that $f(z)$ is analytic in the simply connected domain D containing the disc $D_R(z_0)$. Show that for all positive integers n

$$\left|f^{(n)}(z_0)\right| \le \frac{Mn!}{R^n} \tag{1}$$

where M denotes the maximum value of $|f(z)|$ for z on C, the circumference of the disc.

SOLUTION 3.18

The hypotheses permit the use of the Cauchy integral theorem, (3.6), for the contour $C = \{|z - z_0| = R\}$. On C we have $z = z_0 + Re^{i\vartheta}$ for $0 < \vartheta < 2\pi$ and $dz = iRe^{i\vartheta}d\vartheta$ hence

$$f^{(n)}(z_0) = \frac{n!}{2\pi i}\int_C \frac{f(z)}{(z - z_0)^{n+1}}dz$$

$$= \frac{n!}{2\pi i}\int_0^{2\pi}\frac{f(z_0+Re^{i\vartheta})\,iRe^{i\vartheta}}{R^{n+1}e^{i(n+1)\vartheta}}d\vartheta$$

Then

$$\left|f^{(n)}(z_0)\right| \le \frac{n!}{2\pi R^n}\int_0^{2\pi}\left|f(z_0+Re^{i\vartheta})\right|d\vartheta \le \frac{n!M}{2\pi R^n}\int_0^{2\pi}d\vartheta$$

which leads immediately to (1).

PROBLEM 3.19

Suppose $u = u(r, \vartheta)$ is continuous in the closed disc $D = \{x^2+y^2 \le 1\}$ and u is harmonic on the interior of D. Then show that

$$u(r,\vartheta) = \frac{1}{2\pi}\int_0^{2\pi}\frac{(1-r^2)\,u(1,\varphi)}{1-2r\cos(\vartheta-\varphi)+r^2}d\varphi \tag{1}$$

Equation (1) is known as the *Poisson integral formula* for the disc D and gives the values of a harmonic function $u(x, y)$ inside the disc in terms of its values on the boundary of the disc; i.e., u solves the Dirichlet problem for Laplace's equation on the disc.

SOLUTION 3.19

The hypotheses imply that u is the real part of a complex valued function $f(z)$ that is analytic on $\{|z| < 1\}$. Let C denote the circumference of this disc. Then for z on C, $z = e^{i\varphi}$, $dz = ie^{i\varphi}d\varphi = izd\varphi$. For $z_0 = re^{i\vartheta}$ in D we have by the Cauchy integral formula

$$f(re^{i\vartheta}) = \frac{1}{2\pi i}\int_C\frac{f(z)}{z-z_0}dz = \frac{1}{2\pi}\int_0^{2\pi}\frac{f(z)}{z-z_0}zd\varphi \tag{2}$$

Note that if $z_0 = re^{i\vartheta}$ is inside then $1/z_0{}^* = r^{-1}e^{i\vartheta}$ is outside D and hence

$$0 = \frac{1}{2\pi}\int_0^{2\pi}\frac{f(z)}{z-1/z_0{}^*}zd\varphi$$

$$= \frac{1}{2\pi}\int_0^{2\pi}\frac{f(z)}{z_0{}^*-z^*}z_0{}^*\,d\varphi \tag{3}$$

This integral (3) was obtained by multiplying the numerator and denominator of the previous integrand by $z^*z_0{}^*$ and recalling that $zz^* = 1$. Adding equations (2) and (3) gives

$$f(re^{i\vartheta}) = \frac{1}{2\pi}\int_0^{2\pi}f(z)\left(\frac{z}{z-z_0}+\frac{z_0{}^*}{z_0{}^*-z^*}\right)d\varphi$$

$$= \frac{1}{2\pi}\int_0^{2\pi} \frac{1-|z_0|^2}{|z-z_0|^2} f(e^{i\varphi})\, d\varphi \qquad (4)$$

Since the quotient in the integrand in (4) is a real valued expression, it follows that

$$u(re^{i\vartheta}) = u(r,\vartheta) = \frac{1}{2\pi}\int_0^{2\pi} \left(\frac{1-r^2}{|e^{i\varphi}-re^{i\varphi}|^2}\right) u(1,\varphi)\, d\varphi \qquad (5)$$

Then

$$|e^{i\varphi} - re^{i\varphi}|^2 = |\cos\varphi + i\sin\varphi - r\cos\vartheta - ir\sin\vartheta|^2$$

$$= 1 - 2r\cos(\varphi - \vartheta) + r^2$$

and (1) follows from (5).

PROBLEM 3.20

Suppose $f(z)$ is analytic in the upper half plane $\text{Im}\, z > 0$ and that for positive constants p and M, we have $|z^p f(z)| \le M$ for all z in this half plane. Then show that $u(x, y)$, the real part of $f(z)$, satisfies

$$u(x,y) = \frac{1}{\pi}\int_{-\infty}^{\infty} \frac{y}{(x-\xi)^2 + y^2} u(\xi, 0)\, d\xi \qquad (1)$$

for all (x, y) in the half plane. This is the Poisson integral formula for the half plane and provides a solution to the Dirichlet problem for Laplace's equation on the half plane.

SOLUTION 3.20

Let z_0 denote a fixed but arbitrary point in the upper half plane and let C be the positively oriented closed curve consisting of the straight line from $(-R, 0)$ to $(R, 0)$ together with the circular arc $z = Re^{i\vartheta}$ for $0 \le \vartheta \le \pi$, where R denotes a positive constant larger than $|z_0|$. Then by the Cauchy integral formula

$$f(z_0) = \frac{1}{2\pi i}\int_C \frac{f(z)}{z - z_0} dz \qquad (2)$$

In addition, since $z_0^* = x_0 - iy_0$ lies in the lower half plane outside of C, the Cauchy Goursat theorem implies

$$0 = \frac{1}{2\pi i}\int_C \frac{f(z)}{z - z_0^*} dz \qquad (3)$$

Subtracting (3) from (2) leads to

$$f(z_0) = \frac{1}{2\pi i}\int_C f(z) \left(\frac{1}{z - z_0} - \frac{1}{z - z_0^*}\right) dz$$

$$= \frac{1}{2\pi i}\int_C f(z)\,\frac{z_0 - z_0^*}{(z - z_0)\,(z - z_0^*)}\,dz$$

$$= \frac{1}{2\pi i}\int_{-R}^{R} f(x)\,\frac{2iy_0}{(x - x_0)^2 + (y_0)^2}\,dx + \qquad (4)$$

$$\frac{1}{2\pi i}\int_0^{\pi} f(Re^{i\vartheta})\,\frac{2iy_0}{R^2e^{2i\vartheta} - 2rRe^{i\vartheta}\cos\varphi + r^2}\,iRe^{i\vartheta}\,d\vartheta$$

where $z_0 = (x_0, y_0) = (r, \varphi)$. Then it follows from (4) that

$$\left| f(z_0) - \frac{y_0}{\pi}\int_{-R}^{R}\frac{f(x)}{(x - x_0)^2 + (y_0)^2}\,dx \right| \le \frac{y_0}{\pi}\int_0^{\pi}\frac{M}{R^p}\,\frac{R\,d\vartheta}{\left|R^2e^{2i\vartheta} - 2rRe^{i\vartheta}\cos\varphi + r^2\right|}$$

This holds for all values of $R > |z_0|$ and as R tends to infinity, the integral on the right tends to zero. Then

$$f(z_0) - \frac{y_0}{\pi}\int_{-\infty}^{\infty}\frac{f(x)}{(x - x_0)^2 + (y_0)^2}\,dx = 0 \qquad (5)$$

and, equating real parts in (5) implies

$$u(x_0, y_0) = \frac{y_0}{\pi}\int_{-\infty}^{\infty}\frac{u(x, 0)}{(x - x_0)^2 + (y_0)^2}\,dx$$

This expression, equivalent to (1), gives the value of the harmonic function u at any point (x_0, y_0) in the upper half plane in terms of its values on the real axis.

SUMMARY

The integral of a complex valued function $f(z) = u(x, y) + iv(x, y)$ along a contour C in the complex plane is equivalent to two real planar contour integrals

$$\int_C f(z)\,dz = \int_C u\,dx - v\,dy + i\int_C v\,dx + u\,dy$$

Contour integrals have the following properties for any f and any contour C,

$$\int_{C_1 + C_2} f(z)\,dz = \int_{C_1} f(z)\,dz + \int_{C_2} f(z)\,dz$$

$$\int_{-C} f(z)\,dz = -\int_C f(z)\,dz$$

$$\left|\int_C f(z)\,dz\right| \le \int_C |f(z)|\,|dz| \le ML$$

where

$$M = \max_C |f(z)| \quad \textit{and} \quad L = \int_C |dz| = \text{length of } C$$

If $f(z)$ is not analytic on C then the contour integral must be evaluated by parameterizing C. If $f(z)$ is analytic in a simply connected domain D then it follows from the Cauchy Goursat theorem that for any z_0 and z_1 in D

$$\int_{z_0}^{z_1} f(z)\,dz = F(z_1) - F(z_0)$$

where F is an antiderivative of f. In particular, the integral depends only on the endpoints of the path not on the contour joining them. Additional consequences of the Cauchy Goursat theorem include:

The Cauchy integral formulas (3.5) and (3.6)
The Cauchy estimates (3.7)
The mean value property for analytic and harmonic functions
The maximum principle for analytic functions
The maximum-minimum principle for harmonic functions
The fundamental theorem of algebra

SUPPLEMENTARY PROBLEMS

Evaluate the following integrals by any method:

1. $\int_{-i}^{i} |z|\,dz$ counterclockwise along the half circle $|z| = 1$ from $-i$ to i.
2. $\int_1^{1+i} \sqrt{z}\,dz$ choose the branch where $\sqrt{1} = 1$
3. $\int_{-1-i}^{1+i} \frac{2z-6}{z(z-2)}\,dz$
4. $\int_{-i}^{1} z^{-1}\,dz$
5. $\int_0^{\pi i} z\cos z^2\,dz$
6. $\int_{-i}^{2i} (z^2 + z + 1)\,dz$
7. $\int_C \frac{2i}{z^2+1}\,dz \qquad C = \{|z-1| = 6\}$

8. $\int_C \frac{dz}{z(z+\pi i)}$ $\quad C = \{|z+3i| = 1\}$

9. $\int_C \frac{z^2}{z-i} dz$ $\quad C = \{|z| = 2\}$

10. $\int_C \frac{1}{z^2-1} dz$ $\quad C = \{|z+i| = 5\}$

11. $\int_C \frac{ze^{-z}}{z-\pi i/2} dz$ $\quad C = \{|z| = \pi\}$

12. $\int_C \frac{1}{z^4(z+i)} dz$ $\quad C = \{|z-i| = 3/2\}$

13. $\int_C \frac{z^2 \sin z}{(z-\pi/2)^4} dz$ $\quad C = \{|z-1| = 3/2\}$

14. $\int_C \left(\frac{3}{z+2i} + \frac{4}{z+1}\right) dz$ $\quad C = \{|z| = 4\}$

SOLUTIONS TO SUPPLEMENTARY PROBLEMS

1. $2i$
2. $-2/3\,(1 - i2^{3/4} e^{-i\pi/8})$
3. $\frac{1}{2}\ln 5 + i\,(9\pi/4 + \arctan(1/3))$
4. $\pi i/2$
5. $-\frac{1}{2}\sin\pi^2$
6. $-3/2$
7. 0
8. -2
9. $-2\pi i$
10. 0
11. $i\pi^2$
12. $-2\pi i$
13. $-i\pi^2$
14. $14\pi i$

4

Complex Power Series

In chapter 2 analyticity is characterized in terms of differentiation while chapter 3 examines analyticity from the standpoint of complex contour integrals. Finally, in this chapter we will use complex power series to illustrate yet another aspect of analyticity.

We begin by collecting the essential results relating to power series and then proceed to show that $f(z)$ is analytic at a point z_0 if and only if $f(z)$ has a convergent Taylor series expansion about z_0. The solved problems include several examples of Taylor series expansions. The Laurent series is an extension of the Taylor series concept which permits expansions about singular points.

In the next chapter these power series techniques will be combined to great advantage with the integration results of the previous chapter.

INFINITE SEQUENCES AND SERIES

SEQUENCES

We shall use the notation $\{z_n\}$ to denote an infinite set of complex numbers, $z_0, z_1, \ldots$ We refer to $\{z_n\}$ as a *sequence* and to the individual numbers z_n as the *terms of the sequence.* More precisely, a sequence is a complex valued function whose domain of definition is the nonnegative integers.

CONVERGENCE OF SEQUENCES

The sequence $\{z_n\}$ is said to *converge to the limit L* if and only if for every $\varepsilon > 0$ there exists a positive integer N such that $|z_n - L| < \varepsilon$ for all $n > N$. If a sequence does not converge then it is said to *diverge.*

INFINITE SERIES

An infinite sum of the form

$$\sum_{n=0}^{\infty} z_n, \tag{4.1}$$

obtained by summing the terms of an infinite sequence, is called an *infinite series.* Associated with the infinite series is a sequence, called the *sequence of partial sums* for the series

$$S_m = \sum_{n=0}^{m} z_n. \tag{4.2}$$

The infinite series (4.1) is said to be *convergent to the sum S* if and only if the sequence of partial sums converges to the limit S. If the infinite series is not convergent we say it is *divergent.*

REAL AND COMPLEX SEQUENCES AND SERIES

The following theorem permits us to use all the results relating to convergence of real sequences and series for complex sequences and series.

Theorem 4.1 A sequence of complex numbers is convergent if and only if the sequences of real and imaginary parts of the terms converge as sequences of real numbers. An infinite series of complex numbers is convergent if and only if the series of real and imaginary parts both converge as infinite series of real numbers.

POWER SERIES

INFINITE SERIES OF FUNCTIONS

If the infinite family of functions $\{f_n(z)\}$ has a common domain of definition D then we can form the infinite series

$$\sum_{n=0}^{\infty} f_n(z) = F(z) \tag{4.3}$$

This infinite series defines a new function $F(z)$ whose domain is the set of all z for which the series (4.3) converges.

POWER SERIES

We shall be interested in the series (4.3) in the special case that for each n, $f_n(z) = a_n(z - z_0)^n$, where z_0 and a_n denote specified constants. Then we say (4.3) is a *power series* with *center* at z_0 and *coefficients* a_n.

We can show that a power series necessarily converges on a disc of the form $D_R(z_0) = \{|z - z_0| < R\}$ where exactly one of the following cases applies:

1. $R = 0$ In this case the series converges only at $z = z_0$.
2. $0 < R < \infty$ In this case the series converges at each point of $D_R(z_0)$. In addition, the series converges on the closed disc $\{|z - z_0| \leq r\}$ for every $r < R$. The series diverges outside $D_R(z_0)$.
3. $R = \infty$ In this case the series converges for all z.

The number R is referred to as the *radius of convergence* of the power series.

Theorem 4.2 The radius of convergence of the power series

$$\sum_{n=0}^{\infty} a_n (z - z_0)^n \tag{4.4}$$

is equal to

$$R = \lim_n \left| \frac{a_n}{a_{n+1}} \right| \quad \text{or} \quad R = \frac{1}{\lim_n |a_n|^{1/n}} \quad \text{if the limits exist or equal } \infty.$$

EXAMPLE 4.1

(a) Consider the power series

$$\sum_{n=0}^{\infty} \frac{z^n}{n!}$$

Then

$$\left| \frac{a_n}{a_{n+1}} \right| = n + 1 \to \infty \text{ as } n \text{ tends to infinity; i.e., } R = \infty$$

(b) For the series

$$\sum_{n=0}^{\infty} (-1)^n (z-1)^n$$

we have $|a_n| = 1$ for all n and thus $R = 1$.

(c) For $a_n = n^n$ for all n it is clear that $|a_n|^{1/n} = n$ tends to infinity as n tends to infinity. Then the power series

$$\sum_{n=0}^{\infty} n^n z^n$$

has radius of convergence zero; i.e., $R = 0$.

Theorem 4.3

If the power series (4.4) converges at z_1 different from z_0 then the series also converges at any z whose distance from z_0 is less than $|z_1 - z_0|$. The series converges for all z such that $|z - z_0| < |z_1 - z_0|$.

PROPERTIES OF POWER SERIES

Power series have a number of general properties. The following theorem lists some algebraic properties of power series.

Theorem 4.4

Suppose we have two power series

$$\sum_{n=0}^{\infty} a_n (z - z_0)^n \text{ converging to } F(z) \text{ on } D_R(z_0)$$

$$\sum_{n=0}^{\infty} b_n (z - z_0)^n \text{ converging to } G(z) \text{ on } D_r(z_0)$$

and let ρ denote the smaller of the two positive numbers R and r. Then it follows that

1. $F(z) = G(z)$ for all z in $D_\rho(z_0)$ if and only if $a_n = b_n$ for all n
2. For all constants A and B

$$\sum_{n=0}^{\infty} (Aa_n + Bb_n)(z - z_0)^n = AF(z) + BG(z) \text{ for all } z \text{ in } D_\rho(z_0)$$

3. $$\sum_{n=0}^{\infty} \sum_{m=0}^{n} a_m b_{n-m} (z - z_0)^n = F(z)G(z) \text{ for all } z \text{ in } D_\rho(z_0)$$

DIFFERENTIATION AND INTEGRATION OF POWER SERIES

Power series have very special properties with respect to differentiation and integration that are not typical of infinite series of functions in general. These properties are enumerated in the following theorem.

Theorem 4.5

Suppose that for z_0 given and $R > 0$,

$$\sum_{n=0}^{\infty} a_n (z - z_0)^n \text{ converges on } D_R(z_0) \text{ to the sum } F(z).$$

Then

1. $\sum_{n=1}^{\infty} na_n (z-z_0)^{n-1}$ converges to $F'(z)$ on $D_R(z_0)$

   $\sum_{n=2}^{\infty} n(n-1) a_n (z-z_0)^{n-2}$ converges to $F''(z)$ on $D_R(z_0)$ etc.

2. $\sum_{n=1}^{\infty} \frac{a_n}{n+1} (z-z_0)^{n+1}$ converges to $\int_{z_0}^{z} F(\xi)\,d\xi$ on $D_R(z_0)$

3. $F(z)$ is analytic at z_0 and $F^{(n)}(z_0) = n!a_n$ for all n.

POWER SERIES AS ANALYTIC FUNCTIONS

TAYLOR SERIES

The previous theorem asserts that if a power series with center at z_0 has a positive radius of convergence then the sum of the series is analytic at z_0. The following theorem, known as Taylor theorem, asserts the converse.

Theorem 4.6
Taylor's Theorem

Suppose $F(z)$ is analytic at z_0 and let ξ denote the nearest point to z_0 at which $F(z)$ fails to be analytic. Then

$$\sum_{n=0}^{\infty} a_n (z-z_0)^{n-1} \text{ converges to } F(z) \text{ on } D_R(z_0) \tag{4.4}$$

where

$$a_n = F^{(n)}(z_0)/n! = \frac{1}{2\pi i}\int_C \frac{f(z)}{(z-z_0)^{n+1}}dz \quad \text{for } n = 0, 1, \ldots \tag{4.5}$$

for $C = \{|z-z_0| = r < R\}$ and $R = |z_0 - \xi|$ = distance from z_0 to ξ.

The power series (4.4) with coefficients defined by (4.5) is called the *Taylor series expansion for F about the point* z_0.

Theorem 4.6 and 4.5, together imply that $F(z)$ is analytic at z_0 if and only if the Taylor's series for F at the point z_0 has a positive radius of convergence.

SUBSTITUTION PRINCIPLE

Suppose

$$\sum_{n=0}^{\infty} a_n (z-z_0)^n \text{ converges to } F(z) \text{ on } D_R(z_0) = \{z: |z-z_0| < R\}$$

and that $g(z)$ is analytic on domain D. Then

$$\sum_{n=0}^{\infty} a_n (g(z) - z_0)^n \text{ converges to } F(g(z))$$

at each z in D such that $|g(z) - z_0| < R$.

EXAMPLE 4.2

(a) For $f(z) = e^z$ we have $f^{(n)}(0) = 1$ for all nonnegative integers n. Then the series in Example 4.1(a) is the Taylor series expansion for e^z about the point $z_0 = 0$. As we have seen, this series converges for all z.

(b) The series

$$\sum_{n=0}^{\infty} z^n$$

is seen to converge for $|z| < 1$; e.g. the radius of convergence is equal to 1. Note that if S_m denotes the m-th partial sum then

$$S_m - zS_m = \sum_{n=0}^{m} z^n - \sum_{n=0}^{m} z^{n+1} = 1 - z^{m+1}$$

and

$$S_m = \frac{1-z^{m+1}}{1-z} \qquad \text{for } m = 1, 2, \ldots$$

We can show then that for $|z| < 1$, S_m tends to $(1-z)^{-1}$ as m tends to infinity; i.e.,

$$\sum_{n=0}^{\infty} z^n = \frac{1}{1-z} \qquad \text{for } |z| < 1.$$

We have shown that the power series in this example converges to $f(z) = (1-z)^{-1}$ on $\{|z| < 1\}$. Taylor's theorem asserts that the Taylor series for $f(z)$ about the point $z_0 = 0$ is a series in powers of z also converging to $f(z)$ on this same disc. By the result 1. in Theorem 4.3 this series must be the Taylor series expansion about $z_0 = 0$ for the function $f(z) = (1-z)^{-1}$. Note that $f(z)$ is analytic everywhere except at the point $z = 1$ and that $R = |0 - 1|$ equals the distance from z_0 to $z = 1$.

(c) If we substitute $1 - z$ for z in the previous example we obtain the result

$$\sum_{n=0}^{\infty} (1-z)^n = \sum_{n=0}^{\infty} (-1)^n (z-1)^n = f(1-z) = \frac{1}{z} \text{ for } |1-z| < 1$$

We found the same domain of convergence for this series in Example 4.1(b). By the uniqueness result 1. of Theorem 4.3 this series must be the Taylor series expansion about $z_0 = 1$ for the function $F(z) = 1/z$.

LAURENT SERIES

The Taylor series expansion for $f(z)$ must be centered at a point z_0 where $f(z)$ is analytic and the disc of convergence extends only as far as the nearest point where $f(z)$ fails to be analytic. We can generalize the Taylor series slightly to obtain a power series whose domain of convergence is an *annular region* (i.e., a set of the form $r < |z-z_0| < R$, $r < R$. Then for each z in the annulus $f(z)$ equals the sum of the power series

$$f(z) = \sum_{n=0}^{\infty} a_n (z-z_0)^n + \sum_{n=1}^{\infty} \frac{b_n}{(z-z_0)^n} \tag{4.6}$$

for coefficients given by

$$a_n = \frac{1}{2\pi i}\int_{C_1} \frac{f(z)}{(z-z_0)^{n+1}} \quad \text{for } n = 0, 1, \ldots \tag{4.7}$$

$$b_n = \frac{1}{2\pi i}\int_{C_2} (z-z_0)^{n-1} f(z)\, dz \quad \text{for } n = 1, 2, \ldots \tag{4.8}$$

where $C_1 = \{|z-z_0| = R\}$ and $C_2 = \{|z-z_0| = r\}$ are both traversed in the counterclockwise direction.

The series (4.6) is called the *Laurent series* for $f(z)$ about the point z_0. The integral formulas for the coefficients (4.7) and (4.8) are rarely used to compute the Laurent series for a function.

EXAMPLE 4.3

(a) Consider the function $f(z) = e^{1/z}$. The function $f(1/z) = e^z$ has the everywhere convergent Taylor series shown in Example 4.1(a). Thus by substituting $1/z$ for z in that series we find

$$e^{1/z} = 1 + \frac{1}{z} + \frac{1}{2!z^2} + \frac{1}{3!z^3} + \ldots$$

Since the series in Example 4.1 converges for all $|z| < \infty$, this series converges for $|1/z| < \infty$; i.e., for all $|z| > 0$.

(b) The function $f(z) = z^{-1}(z-1)^{-1}$ has singularities at $z = 0$ and $z = 1$. The function has a Laurent series expansion that is convergent in the annular region $0 < |z| < 1$ given by

$$\frac{1}{z}\frac{1}{z-1} = -\frac{1}{z}\frac{1}{1-z} = -\frac{1}{z}\sum_{n=0}^{\infty} z^n = -\sum_{n=0}^{\infty} z^{n-1}$$

Since the Taylor series used here for $(1-z)^{-1}$ converges for $|z| < 1$, and $1/z$ is analytic for $|z| > 0$, the series for $f(z)$ converges in $0 < |z| < 1$.

(c) We can write another Laurent series for $f(z)$ that converges in the annular region $1 < |z| < \infty$; i.e., in the exterior of the disc $D_1(0)$. To do this we first write a Laurent series for $(z-1)^{-1}$ that converges in the exterior of the disc $D_1(0)$

$$\frac{1}{z-1} = \frac{1}{z}\frac{1}{1-1/z} = \frac{1}{z}\sum_{n=0}^{\infty}\left(\frac{1}{z}\right)^n = \sum_{n=0}^{\infty} z^{-n-1}$$

Since the Taylor series for $(1-z)^{-1}$ converges in $|z| < 1$, the Laurent series for $(1-1/z)^{-1}$ converges in $|z| > 1$. Then since z^{-1} is analytic for $|z| > 0$

$$\frac{1}{z}\frac{1}{z-1} = \frac{1}{z^2}\sum_{n=0}^{\infty}\left(\frac{1}{z}\right)^n = \sum_{n=0}^{\infty} z^{-n-2} \quad \text{for } |z| > 1$$

SOLVED PROBLEMS

Taylor Series

PROBLEM 4.1

Expand $f(z) = \sin z$ in a Taylor series about $z_0 = 0$.

SOLUTION 4.1

Since $f'(z) = \cos z$ and $f''(z) = -\sin z = -f(z)$, it follows that

$$f^{(2m)}(0) = 0 \quad \text{and} \quad f^{(2m+1)}(0) = (-1)^m \quad \text{for } m = 0, 1, \ldots$$

Then

$$\sin z = z - \frac{z^3}{3!} + \frac{z^5}{5!} - \frac{z^7}{7!} + \ldots = \sum_{n=0}^{\infty}\frac{(-1)^n z^{2n+1}}{(2n+1)!} \qquad (1)$$

Since $f(z)$ has no singular points the distance from $z_0 = 0$ to the nearest singular point is infinite; the series has an infinite radius of convergence. Differentiating the series in (1) with respect to z leads to

$$\cos z = \sum_{n=0}^{\infty}\frac{(-1)^n z^{2n}}{(2n)!} \qquad (2)$$

and by part 1 of Theorem 4.5, this series has the same infinite radius of convergence as the series (1).

PROBLEM 4.2

Expand each of the functions

$$f(z) = \frac{1}{1+z}, \quad g(z) = \frac{1}{1+z^2} \quad \text{and} \quad h(z) = \arctan z$$

in a Taylor series about the point $z_0 = 0$.

SOLUTION 4.2

From example 4.2(b) we have

$$\sum_{n=0}^{\infty} z^n = \frac{1}{1-z} \quad \text{for } |z| < 1. \tag{1}$$

Then

$$\frac{1}{1+z} = \frac{1}{1-(-z)} = \sum_{n=0}^{\infty} (-z)^n = 1 - z + z^2 - \ldots \quad \text{for } |z| < 1 \tag{2}$$

Now $g(z) = f(z^2)$, hence

$$\frac{1}{1+z^2} = \sum_{n=0}^{\infty} (-z^2)^n = 1 - z^2 + z^4 - \ldots \quad \text{for } |z| < 1 \tag{3}$$

Finally, since $h(z)$ is an antiderivative of $g(z)$, we have

$$\arctan z = \int_0^z \frac{1}{1+s^2} ds = \sum_{n=0}^{\infty} (-1)^n \frac{z^{2n+1}}{2n+1} \quad \text{for } |z| < 1 \tag{4}$$

PROBLEM 4.3

Expand $f(z) = 1/z$ in a Taylor series about the point $z_0 = 1$. Use this to find similar expansions for $F(z) = \log z$ and $g(z) = 1/z^2$.

SOLUTION 4.3

A Taylor series about $z_0 = 1$ will involve positive powers of $z - 1$. If we write

$$\frac{1}{z} = \frac{1}{1 + (z-1)}$$

then, by substituting $z - 1$ for z in the series (2) of the previous problem, we find

$$\frac{1}{z} = \sum_{n=0}^{\infty} (-1)^n (z-1)^n = 1 - (z-1) + (z-1)^2 - \ldots \quad \text{for } |z-1| < 1$$

Since $F(z) = \log z$ is an antiderivative of $f(z)$, we use Theorem 4.5 to write

$$\log z = \int_1^z \frac{1}{s}ds = \sum_{n=0}^{\infty} (-1)^n \frac{(z-1)^{n+1}}{n+1}$$

$$= (z-1) - \frac{(z-1)^2}{2} + \frac{(z-1)^3}{3} - \ldots$$

This series for $\log z$ converges on $|z-1| < 1$. Finally, we have

$$\frac{1}{z^2} = -\frac{d}{dz}\left(\frac{1}{z}\right) = 1 - 2(z-1) + 3(z-1)^2 - \ldots$$

$$= \sum_{n=1}^{\infty} (-1)^{n+1} n (z-1)^{n-1} \quad \text{for } |z-1| < 1.$$

PROBLEM 4.4

The error function, denoted $\operatorname{erf} z$, is defined as follows

$$\operatorname{erf} z = \frac{2}{\sqrt{\pi}} \int_0^z e^{-\xi^2} d\xi.$$

Find a Taylor's series expansion for $\operatorname{erf} z$ about the point $z_0 = 0$.

SOLUTION 4.4

From Example 4.1(a) and part 2 of Theorem 4.5, we have

$$\operatorname{erf} z = \frac{2}{\sqrt{\pi}} \int_0^z \left(1 - \xi^2 + \frac{\xi^4}{2} - \frac{\xi^6}{6} + \frac{\xi^8}{24} - \ldots\right) d\xi$$

$$= \operatorname{erf} z = \frac{2}{\sqrt{\pi}} \int_0^z \sum_{n=0}^{\infty} \frac{(-1)^n \xi^{2n}}{n!} d\xi = \frac{2}{\sqrt{\pi}} \sum_{n=0}^{\infty} \frac{(-1)^n z^{2n+1}}{n!\,(2n+1)}$$

Since the series for e^z has infinite radius of convergence this series is also everywhere convergent.

PROBLEM 4.5

For fixed real number a, find the Taylor series expansion for $f(z) = (1+z)^a$ about $z_0 = 0$.

SOLUTION 4.5

If a is not an integer then $f(z)$ is multiple valued. In this case we choose the branch of $f(z)$ for which $f(0) = 1$. Now we can easily compute

$$f(0) = 1, \quad f'(0) = a, \quad f''(0) = a(a-1),$$

$$f^{(3)}(0) = a(a-1)(a-2)$$

Then Theorem 4.6 asserts that

$$(1+z)^a = 1 + az + \frac{a(a-1)}{1\cdot 2}z^2 + \frac{a(a-1)(a-2)}{1\cdot 2\cdot 3}z^2 + \frac{a(a-1)(a-2)(a-3)}{1\cdot 2\cdot 3\cdot 4}z^2 \ldots$$

If a is an integer the series terminates at the $(a+1)$st term but if a is not an integer then the series has infinitely many terms. This Taylor series has radius of convergence $R = 1$ since $z_1 = -1$ is a branch point for $f(z)$ and $|z_1 - z_0| = 1$. This result is sometimes referred to as the *binomial theorem* for noninteger exponent.

PROBLEM 4.6

Use the previous result to find a Taylor series expansion about $z_0 = 0$ for the functions

$$f(z) = (1-z^2)^{-1/2} \quad \text{and} \quad F(z) = \arcsin z.$$

SOLUTION 4.6

Using the binomial theorem for $a = -1/2$ we write first

$$(1+z)^{-1/2} = 1 - \frac{1}{2}z + \left(-\frac{1}{2}\right)\left(-\frac{3}{2}\right)\frac{z^2}{2!} + \left(-\frac{1}{2}\right)\left(-\frac{3}{2}\right)\left(-\frac{5}{2}\right)\frac{z^3}{3!} + \ldots$$

Then, substituting $-z^2$ for z,

$$(1+z^2)^{-1/2} = 1 + \frac{1}{2}z + \frac{3}{8}z^4 + \frac{5}{16}z^6 + \ldots \quad \text{for } |z| < 1$$

Then since

$$\arcsin z = \int_0^z (1-t^2)^{-1/2}\,dt$$

we find

$$\arcsin z = z + \frac{1}{2}\frac{z^3}{3} + \frac{3}{8}\frac{z^5}{5} + \frac{5}{16}\frac{z^7}{7} + \ldots \quad \text{for } |z| < 1.$$

Laurent Series

PROBLEM 4.7

Find all possible power series expansions for $f(z) = z^{-1}$ about the point $z_0 = 1$.

SOLUTION 4.7

In Problem 4.3 we found the Taylor series expansion for $f(z)$ about $z_0 = 1$, convergent on the disc $|z-1| < 1$. Now write

$$\frac{1}{z} = \frac{1}{z-1+1} = \frac{1}{z-1}\,\frac{1}{1+\dfrac{1}{z-1}}$$

$$= \frac{1}{z-1}\left(1 - \frac{1}{z-1} + \frac{1}{(z-1)^2} - \frac{1}{(z-1)^3} + \ldots\right)$$

where we have used

$$\frac{1}{1+c} = 1 - c + c^2 - c^3 + \ldots \quad \text{with } c = \frac{1}{z-1} \tag{1}$$

Then

$$\frac{1}{z} = \frac{1}{z-1} - \frac{1}{(z-1)^2} + \frac{1}{(z-1)^3} - \ldots \quad \text{for } \left|\frac{1}{z-1}\right| < 1 \text{ or } |z-1| > 1.$$

This series for $f(z)$ converges in the annular region $1 < |z-1| < \infty$, hence this must be a Laurent series for $f(z)$.

PROBLEM 4.8

Find the following power series expansions about $z_0 = i$ for

$$f(z) = \frac{2z+i}{z(z+i)} = \frac{1}{z} + \frac{1}{z+i},$$

(a) a Taylor's series converging inside a disc centered at $z_0 = i$
(b) a Laurent series converging in an annular region around $z_0 = i$

SOLUTION 4.8

A Taylor series expansion about $z_0 = i$ will involve positive powers of $z - z_0 = z - i$. Thus we begin by writing

$$\frac{1}{z} = \frac{1}{z-i+i} = \frac{1}{i}\frac{1}{\frac{z-1}{i}+i} = \frac{1}{i}\left(1 - \frac{z-i}{i} + \left(\frac{z-i}{i}\right)^2 - \left(\frac{z-i}{i}\right)^3 + \ldots\right)$$

where we have used the result (1) of the previous problem again, this time with $c = (z-i)/i$. This series for $1/z$ is a Taylor series convergent on $|z-i| < 1$ since $R = 1$ equals the distance from $z_0 = i$ to the singular point at $z_1 = 0$. Next we write

$$\frac{1}{z+i} = \frac{1}{2i+z-i} = \frac{1}{2i}\left(\frac{1}{1+\frac{z-i}{2i}}\right)$$

$$= \frac{1}{2i}\left(1 - \frac{z-i}{2i} + \left(\frac{z-i}{2i}\right)^2 - \left(\frac{z-i}{2i}\right)^3 + \ldots\right)$$

This time we applied the result (1) with $c = (z-i)/2i$. This series converges for $|(z-i)/2i| < 1$; i.e., for $|z-i| < 2$. The sum of the two series gives a series for $f(z)$ which converges in the region common to the two discs, namely in the smaller disc, $|z-i|<1$.

A second expansion for $f(z)$ is possible. If we write

$$\frac{1}{z}=\frac{1}{z-i+i}=\frac{1}{z-i}\left(\frac{1}{1+\dfrac{i}{z-i}}\right)=\frac{1}{z-i}\left(1-\frac{i}{z-i}+\left(\frac{i}{z-i}\right)^2-\ldots\right)$$

then it is clear that this is a Laurent series for z^{-1} convergent for $|i/(z-i)|<1$; i.e., for $|z-i|>1$. Adding this to the previously derived series for $(z+i)^{-1}$, we have a series for $f(z)$ that converges in the domain common to the two domains of convergence, namely the annulus $1<|z-i|<2$.

PROBLEM 4.9

Find the following power series expansions about the point $z_0=-i$ for

$$f(z)=\frac{1}{z(z-i)}=\frac{1}{z-i}-\frac{1}{z}$$

(a) a Taylor series converging on $|z+i|<1$
(b) a Laurent series convergent on the annulus $1<|z+i|<2$
(c) a Laurent series that converges on $2<|z+i|$

SOLUTION 4.9

The Taylor series is obtained by writing

$$\frac{1}{z}=\frac{1}{z+i-i}=i\left(\frac{1}{1-\dfrac{z+i}{i}}\right)=i\sum_{n=0}^{\infty}\left(\frac{z+i}{i}\right)^n \quad \text{for } |z+i|<1$$

and

$$\frac{1}{z-i}=\frac{1}{z+i-2i}=\frac{i}{2}\left(\frac{1}{1-\dfrac{z+i}{2i}}\right)=\frac{i}{2}\sum_{n=0}^{\infty}\left(\frac{z+i}{2i}\right)^n \quad \text{for } |z+i|<2$$

Then subtracting the first of these series from the second gives a Taylor series for $f(z)$ that converges on $|z+i|<1$, the common domain of convergence of the two series.

If we write

$$\frac{1}{z}=\frac{1}{z+i-i}=\frac{1}{z+i}\frac{1}{1-\dfrac{i}{z+i}}=\frac{1}{z+i}\sum_{n=0}^{\infty}\left(\frac{i}{z+i}\right)^n$$

then this series converges for $|i/(z+i)|<1$; i.e., for $|z+i|>1$. If we subtract this series from the previously obtained series for $(z-i)^{-1}$ then the result is a series which converges on $1<|z+i|<2$, the common domain of convergence for the two series.

Finally, using the series just derived for z^{-1} together with the series

$$\frac{1}{z-i} = \frac{1}{z+i-2i} = \frac{1}{z+i}\frac{1}{1-\frac{2i}{z+i}} = \frac{1}{z+i}\sum_{n=0}^{\infty}\left(\frac{2i}{z+i}\right)^n$$

which converges for $|2i/(z+i)| < 1$ (for $2 < |z+i|$), we obtain a series for $f(z)$ which converges on the exterior of the disc $|z+i| < 2$.

Properties of Power Series

PROBLEM 4.10

Suppose that if $f(z)$ is analytic inside the circle $C_0 = \{|z-z_0| = R\}$. Then show that for each z such that $|z-z_0| < R$,

$$S_m(z) = \sum_{n=0}^{m} \frac{f^{(n)}(z_0)}{n!}(z-z_0)^n$$

tends to $f(z)$ as m tends to infinity.

SOLUTION 4.10

Fix an arbitrary point z inside C_0. Then $|z-z_0| = r < R$ and for r_1 such that $r < r_1 < R$ let C_1 denote the circle of radius r_1 centered at z_0. Thus C_1 is inside C_0 and z is inside C_1. Then the Cauchy integral formula gives

$$f(z) = \frac{1}{2\pi i}\int_{C_1}\frac{f(z)}{s-z}ds \tag{1}$$

Now write

$$\frac{1}{s-z} = \frac{1}{s-z_0-(z-z_0)} = \frac{1}{s-z_0}\frac{1}{1-c}$$

$$= \frac{1}{s-z_0}\left(1+c+c^2+\ldots+c^{m-1}+\frac{c^m}{1-c}\right) \tag{2}$$

where

$$c = \frac{z-z_0}{s-z_0} \neq 1. \text{ (See also Example 4.2(b))}$$

Substituting (2) into (1) and using the version (3.6) of the Cauchy integral formula leads to

$$\frac{1}{2\pi i}\int_{C_1}\frac{f(z)}{s-z}ds = f(z_0) + f'(z_0)(z-z_0) + \frac{1}{2!}f''(z_0)(z-z_0)^2 + \ldots$$

$$\ldots + \frac{1}{(n-1)!}f^{(m-1)}(z_0)(z-z_0)^{m-1} + R_m(z)$$

where

$$R_m(z) = \frac{(z-z_0)^m}{2\pi i}\int_{C_1}\frac{f(s)\,ds}{(s-z)\,(s-z_0)^m}$$

Note that for each positive integer m, $R_m(z)$ equals the difference between $f(z)$ and $S_m(z)$. Thus the convergence will be proved it we can show that $R_m(z)$ tends to zero with increasing m. By construction we have

$$|z-z_0| = r < r_1 = |s-z_0| < R$$

and the triangle inequality implies $|s-z_0| \le |s-z| + |z-z_0|$. Therefore, $|s-z| \ge r_1 - r > 0$ for all s on C_1. It follows that if M denotes the maximum value of $|f(s)|$ on C_1 then

$$|R_m(z)| \le \frac{r^m}{2\pi}\frac{M}{(r_1-r)\,r_1^m}2\pi r_1 = \frac{Mr_1}{r_1-r}\left(\frac{r}{r_1}\right)^m \tag{3}$$

Since $r < r_1$, (3) implies that $|R_m(z)|$ tends to zero as m tends to infinity. This proves that $S_m(z)$, the partial sums of the Taylor series for $f(z)$, tend to $f(z)$ with increasing m for every z inside C_0.

PROBLEM 4.11

Suppose that the power series

$$\sum_{n=0}^{\infty} a_n(z-z_0)^n \tag{1}$$

has a radius of convergence $R > 0$ and let r be any positive number such that $r < R$. Then show that for any $\varepsilon > 0$ there is an integer m such that

$$\left|\sum_{n=m}^{\infty} a_n(z-z_0)^n\right| < \varepsilon \quad \text{for all } z \text{ such that } |z-z_0| \le r. \tag{2}$$

We say then that the series (1) converges *absolutely and uniformly* on $D_r(z_0)$.

SOLUTION 4.11

Let C_0 denote the circle $\{|z-z_0| = R\}$ and for fixed, r, $0 < r < R$, let C_1 denote the circle $\{|z-z_0| = r_1\}$ for $r < r_1 < R$. Then C_1 is inside C_0 and the radius of C_1 exceeds r; i.e. $\alpha = r/r_1$ satisfies $0 < \alpha < 1$. By hypothesis, the series (1) converges for all z on C_1 and thus the modulus of the nth term of the series must tend to zero as n tends to infinity. In particular, there exists an N_1 such that $n > N_1$ implies

$$|a_n(z-z_0)^n| = |a_n|\,(r_1)^n < 1-\alpha \tag{3}$$

In addition, since $0 < \alpha < 1$, for any $\varepsilon > 0$ there exists an N_2 such that

$$\alpha^n < \varepsilon \quad \text{for } n > N_2 \tag{4}$$

Choosing N to be the larger of the two numbers N_1, N_2 it follows that (3) and (4) both hold when n is larger than N. Then for $m > N$ and $k > 0$ and any z that is inside C_1

$$\left|\sum_{n=m}^{m+k} a_n (z-z_0)^n\right| \leq \sum_{n=m}^{m+k} |a_n| |z-z_0|^n \leq \sum_{n=m}^{m+k} |a_n| r^n$$

$$\leq \sum_{n=m}^{m+k} |a_n| (r_1)^n \alpha^n \quad (\text{i.e., } r = r_1\alpha)$$

$$\leq (1-\alpha) \sum_{n=m}^{m+k} \alpha^n = \alpha^m - \alpha^{m+k+1} < \varepsilon(1-\alpha^{k+1})$$

Letting k tend to infinity here gives

$$\left|\sum_{n=m}^{\infty} a_n (z-z_0)^n\right| < \varepsilon.$$

PROBLEM 4.12

Suppose the power series (1) of the previous problem converges on $D_R(z_0)$ to the sum $F(z)$. Then show that $F(z)$ is *uniformly continuous* on the *closed disc* $D = \{|z-z_0| \leq r\}$ for every $r < R$; i.e., for every $\varepsilon > 0$ there exists a $\delta > 0$ such that $|F(z_1) - F(z_2)| < \varepsilon$ for all z_1, z_2 in D such that $|z_1 - z_2| < \delta$.

SOLUTION 4.12

Let r, $0 < r < R$, and $\varepsilon > 0$ be given. We have assumed that the power series (1) of the previous problem converges on $D_R(z_0)$ to the sum $F(z)$. Hence by the result of the previous problem, the series converges absolutely and uniformly of $F(z)$ on $D = \{|z-z_0| \leq r\}$. That is, there exists an integer N such that for any z_1 and z_2 in D

$$\text{for } z = z_1, z_2 \quad \left|F(z) - \sum_{n=0}^{m} a_n (z-z_0)^n\right| < \frac{\varepsilon}{3} \quad \text{for } m > N \tag{1}$$

In addition, since

$$P_m(z) = \sum_{n=0}^{m} a_n (z-z_0)^n$$

is a polynomial in z, it is everywhere continuous and hence there exists a $\delta > 0$ such that

$$|P_m(z_1) - P_m(z_2)| < \frac{\varepsilon}{3} \quad \text{for} \quad |z_1 - z_2| < \delta \tag{2}$$

It follows that for the chosen ε and this δ,

$$\begin{aligned}
&|F(z_1) - F(z_2)| \\
&= |F(z_1) - P_m(z_1) + P_m(z_1) - P_m(z_2) + P_m(z_2) - F(z_2)| \\
&\leq |F(z_1) - P_m(z_1)| + |P_m(z_1) - P_m(z_2)| + |P_m(z_2) - F(z_2)| \\
&\leq \frac{\varepsilon}{3} + \frac{\varepsilon}{3} + \frac{\varepsilon}{3} = \varepsilon \quad \text{for} \quad |z_1 - z_2| < \delta
\end{aligned}$$

Thus the sum of the series must be uniformly continuous on every closed disc inside $D_R(z_0)$.

PROBLEM 4.13

Suppose the power series (1) of Problem 4.11 has positive radius of convergence R. Then for any contour C lying inside the disc $D_R(z_0)$ show that

$$\int_C \sum_{n=0}^{\infty} a_n (z - z_0)^n dz = \sum_{n=0}^{\infty} \int_C a_n (z - z_0)^n dz$$

SOLUTION 4.13

Let C be a contour satisfying $|z - z_0| \leq r < R$ for all z on C. Note that for each n, $f_n(z) = a_n(z - z_0)^n$ is everywhere continuous and hence the integral of f_n on C exists. Similarly, the sum $F(z)$ of the series is continuous inside $D_R(z_0)$ by the result of the previous problem and thus the integral of F over C exists. Now

$$\begin{aligned}
\int_C F(z)\,dz &= \int_C \left(F(z) - \sum_{n=0}^{m} f_n(z) + \sum_{n=0}^{m} f_n(z) \right) dz \\
&= \int_C \left(F(z) - \sum_{n=0}^{m} f_n(z) \right) dz + \sum_{n=0}^{m} \int_C f_n(z)\,dz
\end{aligned}$$

hence

$$\left| \int_C F(z)\,dz - \sum_{n=0}^{m} \int_C f_n(z)\,dz \right| \leq \int_C \left| F(z) - \sum_{n=0}^{m} f_n(z) \right| |dz|$$

In Problem 4.11we showed that for any $\varepsilon > 0$ there exists an N such that for all z such that $|z - z_0| \leq r < R$

$$\left| F(z) - \sum_{n=0}^{m} f_n(z) \right| < \varepsilon \quad \text{for} \quad m > N$$

Then for $m > N$,

$$\left| \int_C F(z)\,dz - \sum_{n=0}^{m} \int_C f_n(z)\,dz \right| \le \varepsilon L \quad (L = \text{length of } C);$$

i.e., the sum of the series of integrals equals the integral of the sum of the series.

PROBLEM 4.14

Suppose the power series (1) of Problem 4.11 converges to the sum $F(z)$ on $D_R(z_0)$. Then show that $F(z)$ is analytic on $D_R(z_0)$.

SOLUTION 4.14

In Problem 4.12 we showed that the Taylor series converges to a sum $F(z)$ that is uniformly continuous on $D_r(z_0)$. We show now that $F(z)$ is, in fact, analytic in $D_r(z_0)$. Let C denote an arbitrary simple closed contour inside $D_r(z_0)$. Then by the result of the previous problem

$$\int_C F(z)\,dz = \sum_{n=0}^{\infty} \int_C a_n (z - z_0)^n dz \qquad (1)$$

The Cauchy Goursat theorem implies that each of the integrals in the sum on the right of (1) equals zero. Then the integral of F around any simple closed contour C inside $D_r(z_0)$ vanishes and Morera's theorem, Theorem 3.12, implies that $F(z)$ must be analytic in $D_r(z_0)$.

PROBLEM 4.15

Suppose the power series (1) of Problem 4.11 converges to the sum $F(z)$ on $D_r(z_0)$. Then show that the series

$$\sum_{n=1}^{\infty} n a_n (z - z_0)^{n-1}$$

converges absolutely and uniformly to $F'(z)$ for all z such that $|z - z_0| \le r_1 < r$.

SOLUTION 4.15

Let z be an arbitrary point in the interior of $D_r(z_0)$. Then there is a simple closed contour C lying inside $D_r(z_0)$ such that z is inside C. Then since $F(z)$ is analytic on and inside C by the result of the previous problem, we have

$$f'(z) = \frac{1}{2\pi i} \int_C \frac{F(s)}{(s - z)^2} ds$$

$$= \frac{1}{2\pi i}\int_C \frac{1}{(s-z)^2} \sum_{n=0}^{\infty} a_n (s-z_0)^n ds$$

$$= \sum_{n=0}^{\infty} a_n \frac{1}{2\pi i}\int_C \frac{(s-z_0)^n}{(s-z)^2} ds$$

The interchange of the order of summation and integration here can be justified by the fact that the power series converges uniformly on the contour C. Now we use the Cauchy integral formula to conclude that for each n,

$$\frac{1}{2\pi i}\int_C \frac{(s-z_0)^n}{(s-z)^2} ds = n(z-z_0)^{n-1}$$

and thus

$$F'(z) = \sum_{n=0}^{\infty} a_n n (z-z_0)^{n-1} = \sum_{n=0}^{\infty} f_n'(z)$$

To see that the convergence is uniform on every closed disc $\{|z-z_0| \le r_1 < r\}$ inside $D_r(z_0)$ write

$$\left| F'(z) - \sum_{n=0}^{m-1} f_n'(z) \right| = \left| \sum_{n=m}^{\infty} f_n'(z) \right| \tag{1}$$

Let C_1 denote the circle $|s-z_0| = r_1 < r$ and let z lie inside C_1. Then

$$\left| \sum_{n=m}^{\infty} f_n'(z) \right| = \left| \sum_{n=m}^{\infty} a_n \frac{1}{2\pi i}\int_C \frac{(s-z_0)^n}{(s-z)^2} ds \right| \tag{2}$$

and since for each $\varepsilon > 0$ there exists an N such that

$$\left| \sum_{n=m}^{\infty} a_n (s-z_0)^n \right| \le \varepsilon \quad \text{for } m > N$$

it follows that

$$\left| \sum_{n=m}^{\infty} a_n \frac{1}{2\pi i}\int_C \frac{(s-z_0)^n}{(s-z)^2} ds \right| \le \frac{1}{2\pi} \frac{\varepsilon 2\pi r_1}{\rho^2} = \varepsilon r_1/\rho^2 \tag{3}$$

where ρ denotes the minimum value of $|z-s|$ for s on C_1. Since z lies inside C_1, ρ is positive. Using (3) in (2) implies that for any z inside C_1,

$$\left| \sum_{n=m}^{\infty} f_n'(z) \right|$$

can be made arbitrarily small by choosing m large. This implies the uniform convergence of the differentiated series on any closed disc inside $D_r(z_0)$.

SUMMARY

We have now three equivalent characterizations of analyticity:

(a) the complex valued function $f(z)$ is analytic throughout a domain D if the first derivative $f'(z)$ exists at each point of D

(b) $f(z)$ is analytic in the simply connected domain D if and only if the integral of f around any closed curve C in D is equal to zero.

(c) $f(z)$ is analytic in the simply connected domain D if and only if f has a convergent Taylor series expansion about each point of D

Suppose $F(z)$ is analytic at $z = z_0$ and that $R > 0$ denotes the distance from z_0 to the nearest singular point for F (if $F(z)$ is entire then R is infinite). Then the Taylor series for F about the point z_0

$$\sum_{n=0}^{\infty} a_n (z-z_0)^n = \sum_{n=0}^{\infty} \frac{F^{(n)}(z_0)}{n!} (z-z_0)^n$$

converges on $D_R(z_0)$ to the sum $F(z)$. This series may be differentiated or integrated term by term any number of times and the resulting series converges (on the same disc) to the appropriate derivative or antiderivative of $F(z)$. In addition, if $g(z)$ is analytic on domain D then

$$\sum_{n=0}^{\infty} a_n (g(z) - z_0)^n \text{ converges to } F(g(z))$$

at each z in D such that $|g(z) - z_0| < R$.

If z_0 is a singular point for $F(z)$ but F is analytic in the annular domain $r < |z - z_0| < R$, the F has a Laurent series expansion of the form

$$\sum_{n=0}^{\infty} a_n (z-z_0)^n + \sum_{n=1}^{\infty} \frac{b_n}{(z-z_0)^n}$$

converging to F on the annulus. Here $r \geq 0$ and $R \leq \infty$. The Laurent series may be differentiated or integrated term by term and the resulting series converges on the same annulus to the appropriate derivative or antiderivative of the sum of the original series. Substitution may also be applied just as in the case of the Taylor series.

The formula for the sum of a geometric series

$$\frac{1}{1+c} = \sum_{n=0}^{\infty} (-1)^n c^n \quad \text{for } |c| < 1$$

implies, via the substitution principle, that for arbitrary complex numbers z_0 *and* a*, we have*

$$\frac{1}{z+a} = \frac{1}{a+z_0+z-z_0} = \frac{1}{a+z_0} \frac{1}{1+\dfrac{z-z_0}{a+z_0}} = \frac{1}{a+z_0} \sum_{n=0}^{\infty} (-1)^n \left(\frac{z-z_0}{a+z_0}\right)^n$$

converging on the disc, $|z-z_0| < |z_0+a|$. *Similarly, for the exterior domain,* $|z-z_0| > |z_0+a|$ *we have*

$$\frac{1}{z+a} = \frac{1}{a+z_0+z-z_0} = \frac{1}{z+z_0} \frac{1}{1+\dfrac{a+z_0}{z-z_0}} = \frac{1}{z-z_0} \sum_{n=0}^{\infty} (-1)^n \left(\frac{a+z_0}{z-z_0}\right)^n$$

These expressions can often be used to build Taylor or Laurent series expansions about the point z_0 *for rational functions. Additional substitutions together with differentiation or integration of series can then be employed to generate additional series.*

SUPPLEMENTARY PROBLEMS

Find a power series expansion for $f(z)$ converging in the indicated region:

1. $f(z) = (z+3)^{-1}$ converging in $|z| < 3$.

2. $f(z) = (z+3)^{-1}$ converging in $|z| > 3$.

3. $f(z) = 1/(z^2+z-2)$ converging in $1 < |z-2| < 4$.

4. $f(z) = 1/(z^2+z-2)$ converging in $4 < |z-2|$.

5. $f(z) = 1/(z^3-2z^2+z)$ converging in $0 < |z-1| < 1$.

Expand $f(z)$ about z_0 and indicate where the series converges:

6. $f(z) = \dfrac{\cos(z^2) - 1}{z^2}$ at $z_0 = 0$

7. $f(z) = \dfrac{e^z - z - 1}{z^2}$ at $z_0 = 0$

8. $f(z) = (z+1)^{-1}$ at $z_0 = i$

SOLUTIONS TO SUPPLEMENTARY PROBLEMS

1. $\dfrac{1}{3}\sum_{n=0}^{\infty} (-1)^n (z/3)^n$

2. $\sum_{n=0}^{\infty} (-1)^n 3^n / z^{n+1}$

3. $\dfrac{1}{12}\sum_{n=0}^{\infty} (-1)^n \left(4(z-2)^{-n-1} - \left(\dfrac{z-2}{4}\right)^n\right)$

4. $\dfrac{1}{3}\sum_{n=0}^{\infty} (-1)^n \dfrac{1-4^n}{(z-2)^{n+1}}$

5. $\sum_{n=0}^{\infty} (-1)^n (z-1)^{n-2}$

6. $\sum_{n=1}^{\infty} \dfrac{(-1)^n}{(2n)!} z^{4n-2}$ all z

7. $\displaystyle\sum_{n=2}^{\infty} \frac{z^{n-2}}{n!}$ for all z

8. $\displaystyle\frac{1}{1+i}\sum_{n=0}^{\infty} (-1)^n \left(\frac{z-i}{1+i}\right)^n$ converges for $|z-i| < \sqrt{2}$

$\displaystyle\frac{1}{z-i}\sum_{n=0}^{\infty} (-1)^n \left(\frac{z-i}{1+i}\right)^{-n}$ converges for $|z-i| > \sqrt{2}$

5

The Residue Theorem and the Argument Principle

In this chapter we combine the series expansion techniques of the previous chapter with the integration results of Chapter 3 to obtain two very powerful analytic tools: the residue theorem and the argument principle.

The residue theorem is an efficient means for evaluating certain definite integrals. It is especially valuable in cases where standard methods fail due to the lack of an explicit antiderivative. Even in situations where an antiderivative is available, the definite integral is generally more easily evaluated by the residue theorem.

The argument principle is a consequence of the residue theorem that allows the zeros of an analytic function to be located without requiring that the zeros be calculated. This ability will be exploited later in connection with studies of stability of linear systems.

We begin the chapter by introducing some more precise terminology for the zeros and singular points of analytic functions and then proceed to develop the residue theorem and the principle of the argument.

SINGULARITIES AND ZEROS OF ANALYTIC FUNCTIONS

SINGULAR POINTS

An *entire function* is a complex valued function that is analytic at all points of the complex plane. On the other hand, we have functions like $f(z) = z^*$ that are nowhere analytic. Between these two extremes there are functions that are analytic at some points but are not analytic everywhere. A point z_0 is called a *singularity* or *singular point* for the complex valued function $f(z)$ if $f(z)$ fails to be analytic at z_0 but every neighborhood of z_0 contains at least one point where $f(z)$ is analytic. A more precise classification of singular points is possible and necessary.

ISOLATED SINGULARITIES

The singular point z_0 is said to be an *isolated singularity* for $f(z)$ if for some $r > 0$ $f(z)$ is analytic in the set $D_r'(z_0) = \{0 < |z - z_0| < r\}$ but $f(z)$ is not analytic at z_0. The set $D_r'(z_0)$ is called a *deleted neighborhood* of z_0.

EXAMPLE 5.1

The point $z_0 = 0$ is a singular point for each of the functions:

$$f_1(z) = 1/z, \quad f_2(z) = 1/z^3, \quad f_3(z) = 1/\sin(\pi/z)$$

$$f_4(z) = \sqrt{z}, \quad f_5(z) = e^{1/z}$$

The point is an isolated singularity for functions $f_1(z)$, $f_2(z)$ and $f_5(z)$ since each of these functions is analytic everywhere except at $z_0 = 0$. The function $f_4(z)$ denotes a branch of the double valued square root function and thus f_4 fails to be analytic at any point on the branch cut. Since the branch cut is a ray originating at $z_0 = 0$, it is clear that there is no deleted neighborhood of z_0 in which f_4 is analytic and thus z_0 is not an isolated singularity for f_4. The function $f_3(z)$ fails to be analytic at z_0, where $\sin(\pi/z)$ fails to be analytic, and at the infinite set of points where $\sin(\pi/z)$ is zero, namely the points $z_n = 1/n$ for n a nonzero integer. Each of the points z_n is an isolated singular point since each of these is separated from its nearest neighbors z_{n-1}, z_{n+1} by a positive distance. However, every neighborhood of z_0 contains singular points z_n so z_0 is not isolated.

CLASSIFICATION OF ISOLATED SINGULARITIES

Suppose z_0 is an isolated singularity for the function $f(z)$. This isolated singular point can be further classified by examining the Laurent series expansion that converges to $f(z)$ in a deleted neighborhood of z_0. Suppose the following Laurent series for $f(z)$ converges in the deleted neighborhood, $0 < |z - z_0| < r$, for some $r > 0$,

$$f(z) = \sum_{n=0}^{\infty} a_n (z - z_0)^n + \sum_{n=1}^{\infty} b_n (z - z_0)^{-n} \tag{5.1}$$

We say that:

(a) z_0 is a *removable singularity* if $b_n = 0$ for all n

(b) z_0 is a *pole of order m* if b_m is not zero but $b_n = 0$ for $n > m$

(c) z_0 is an *essential singularity* if b_n is nonzero for infinitely many n

It will be convenient to have alternative equivalent characterizations of the various isolated singularities.

Theorem 5.1 Suppose z_0 is an isolated singularity of the complex valued function $f(z)$.

1. z_0 is a removable singularity if and only if $f(z)$ tends to a finite limit as z approaches z_0.
2. z_0 is a pole if and only if $f(z)$ tends to infinity as z approaches z_0.
3. z_0 is a pole of order m if and only if $g(z) = (z-z_0)^m f(z)$ tends to a finite nonzero limiting value as z approaches z_0.
4. z_0 is an essential singularity if and only if $f(z)$ tends to no limit, finite or infinite, as z approaches z_0.

EXAMPLE 5.2

(a) $f(z) = (\sin z)/z$ has a removable singularity at $z_0 = 0$. To see this write

$$\sin z = z - \frac{z^3}{3!} + \frac{z^5}{5!} - \ldots \quad \text{for all } z$$

Then

$$\frac{\sin z}{z} = 1 - \frac{z^2}{3!} + \frac{z^4}{5!} - \ldots \quad \text{for } |z| > 0$$

This last series is the Laurent series expansion for $f(z)$, valid in the deleted neighborhood $|z| > 0$. Since the series contains no negative powers of z it follows that $z_0 = 0$ is a removable singularity for $f(z)$. It is also evident from the series that $f(z)$ tends to the value 1 as z approaches zero. Thus part 1 of Theorem 5.1 provides an alternative means of seeing that $z_0 = 0$ is a removable singularity.

(b) $g(z) = (1-\cos z)/z^3$ has a pole of order one at $z_0 = 0$. To see this write

$$g(z) = \frac{1 - (1 - \frac{z^2}{2!} + \frac{z^4}{4!} + \ldots)}{z^3} = \frac{1}{2!\,z} - \frac{z}{4!} + \frac{z^3}{6!} - \ldots \quad \text{for } |z| > 0$$

Then since $b_1 = 1/2!$ and $b_n = 0$ for $n > 1$ we have a pole type singularity for $g(z)$ at z_0. A pole of order one is usually called a *simple pole.*

(c) The origin is a singular point for the function $F(z) = \exp(1/z^2)$. The expansion

$$e^{1/z^2} = 1 + \frac{1}{z^2} + \frac{1}{2!z^4} + \dots$$

converges for all z different from zero and represents the Laurent expansion for $F(z)$ about $z_0 = 0$. Since this series contains infinitely many terms having negative powers of z, $z_0 = 0$ must be an essential singularity. This can also be seen by noting that as z tends to zero along the real axis we have $z = x$ so that

$$e^{1/z^2} = e^{1/x^2} \to \infty \quad \text{as } x \text{ approaches } 0.$$

However, as z tends to zero along the imaginary axis we have $z = iy$ and

$$e^{1/z^2} = e^{-1/y^2} \to 0 \quad \text{as } y \text{ approaches } 0.$$

Thus $F(z)$ tends to no limit as z approaches the essential singularity at $z = 0$.

ZEROS OF FUNCTIONS

It will also be convenient to have a classification of the zeros of analytic functions. We say that z_0 is a *zero of order m* (or of *multiplicity m*) for the analytic function $f(z)$ if $f(z) = (z - z_0)^m g(z)$ where $g(z_0)$ is not equal to zero.

EXAMPLE 5.3

The function $f(z) = z^3(e^z - 1)$ has a zero of order 4 at $z_0 = 0$ since

$$f(z) = z^3\left(z + \frac{z^2}{2!} + \frac{z^3}{3!} + \dots\right) = z^4\left(1 + \frac{z}{2!} + \frac{z^2}{3!} + \dots\right) = z^4 g(z)$$

MEROMORPHIC FUNCTIONS

We have already considered *rational functions* which are defined as quotients of two polynomials having no common factors. More generally we have the class of *meromorphic functions* $f(z) = F(z)/G(z)$, which are quotients of two entire functions F and G. Then the zeros of F are the zeros of f and the zeros of G are the poles of f where z_0 is a pole of order m for f if z_0 is a zero for G of multiplicity m.

THE RESIDUE THEOREM

RESIDUES

Suppose that $f(z)$ is analytic in a deleted neighborhood $D_r'(z_0)$ of an isolated singular point z_0 and let

$$f(z) = \sum_{n=0}^{\infty} a_n (z-z_0)^n + \sum_{n=1}^{\infty} \frac{b_n}{(z-z_0)^n} \tag{5.2}$$

denote the Laurent series for $f(z)$ centered at z_0 and converging to f in $D_r'(z_0)$. Then we define the *residue of* $f(z)$ *at* z_0, $\text{Res}f(z_0)$, to be the coefficient b_1 in (5.2). The importance of the residue lies in the following theorem.

Theorem 5.2 The Residue Theorem Suppose that $f(z)$ is analytic on and inside a simple closed contour C except at isolated singular points $z_1, \ldots, z_m$, lying inside C. Then

$$\int_C f(z)\,dz = 2\pi i \sum_{j=1}^{m} \text{Res}f(z_j)$$

Theorem 5.2 provides a powerful method for evaluating integrals but to take advantage of the theorem we need efficient ways of evaluating residues.

EVALUATING RESIDUES

The residue of $f(z)$ at the isolated singular point z_0 may always be found from the Laurent series (5.2) converging to f in a deleted neighborhood of z_0. If z_0 is an essential singularity then this is the only way to find the residue. If z_0 is not an essential singularity then the residue can be found more simply.

Removable Singularity If z_0 is a removable singularity for $f(z)$, then $b_n = 0$ for all n and the residue of $f(z)$ at z_0 is zero.

Simple Pole If z_0 is a simple pole for $f(z)$ then it is clear from (5.2) that

$$b_1 = \lim_{z \to z_0} (z-z_0) f(z) = \text{Res}f(z_0) \tag{5.3}$$

For example, (5.3) implies

$$\text{if } f(z) = \frac{g(z)}{z-z_0} \text{ for } g(z_0) \neq 0, \text{ then } \text{Res}f(z_0) = g(z_0) \tag{5.4}$$

$$\text{if } f(z) = p(z)/q(z) \text{ where } q(z_0) = 0, p(z_0) \neq 0, \tag{5.5}$$

$q'(z_0) \neq 0$ then $\text{Res} f(z_0) = p(z)/q'(z)$

Pole of order m If z_0 is a pole of order m for $f(z)$ then

$$b_1 = \frac{1}{(m-1)!} g^{(m-1)}(z_0) = \text{Res} f(z_0) \tag{5.6}$$

where $g(z) = (z - z_0)^m f(z)$ is analytic in a neighborhood of z_0 and $g(z_0) \neq 0$.

EXAMPLE 5.4

(a) The function

$$f(z) = \frac{1}{z^2+1} = \frac{1}{(z+i)(z-i)}$$

has simple poles at $z = i, -i$. Then $\text{Res} f(i)$ can be evaluated by either (5.3) or by (5.4) with $g(z) = 1/(z+i)$ to find $\text{Res} f(i) = 1/2i$.

(b) The meromorphic function

$$f(z) = \frac{\sin z}{\cosh z}$$

has isolated singularities at the zeros of $\cosh z$; i.e., at $z_n = (2n+1)i\pi/2$, for integer values of n. Since $\sin z_n$ and $\sinh z_n$ are not zero, these are simple poles and we may apply (5.5) to find

$$\text{Res} f(z_n) = \frac{\sin z_n}{\sinh z_n} = (-1)^n \sinh (2n+1)\pi/2$$

(c) The function $f(z) = z^{-3} e^z$ has a pole of order 3 at $z_0 = 0$. Then $g(z) = z^3 f(z) = e^z$ and by (5.6)

$$\text{Res} f(0) = \frac{1}{2} g''(0) = \frac{1}{2}$$

(d) The function $f(z) = z^{-4}(\sin z - z)$ has a pole at $z = 0$ but the order of this pole is not immediately evident. If we write

$$\sin z - z = -\frac{z^3}{6} + \frac{z^5}{120} - \ldots$$

then

$$f(z) = -\frac{1}{6z} + \frac{z}{120} - \ldots$$

from which it follows by definition that $\text{Res} f(0) = -1/6$.

(e) The function $f(z) = \sqrt{z}/(z+1)^2$ has a pole of order two at $z = -1$ provided the branch cut for $\sqrt{z}$ is chosen to lie somewhere other than the negative real axis. If we put the branch cut on the positive real axis so that $0 < \vartheta \leq 2\pi$, then

$$\operatorname{Re} f(-1) = g'(e^{i\pi}) = \left.\frac{1}{2\sqrt{z}}\right|_{z = e^{i\pi}} = \frac{1}{2}e^{-i\pi/2} = -i/2.$$

EVALUATING INTEGRALS BY THE RESIDUE THEOREM

The residue theorem can be used to evaluate certain real definite integrals whose value cannot be determined by the fundamental theorem of calculus. For certain other integrals where the fundamental theorem applies, the residue theorem may involve less effort. Definite integrals where the residue theorem may be useful include:

1. Integrals of the form $\int_0^{2\pi} F(\cos\vartheta, \sin\vartheta)\, d\vartheta$.

   For z on the circle $C = \{|z| = 1\}$ we have $\cos\vartheta = (z + 1/z)/2$, $\sin\vartheta = (z - 1/z)/2i$ and $dz = d\vartheta/iz$. Then

$$\int_0^{2\pi} F(\cos\vartheta, \sin\vartheta)\, d\vartheta = -i\int_C F((z + 1/z)/2, (z - 1/z)/2i)\, z^{-1} dz$$

2. Integrals of the form $\int_{-\infty}^{\infty} P(x)/Q(x)\, dx$

   Here we must have polynomials P and Q with degree Q - degree $P \geq 2$. Then we say $R(x) = P(x)/Q(x)$ has a zero of order two at infinity and this is enough to guarantee the improper integral converges if there is no pole on the real axis. The residue theorem is then just a method for evaluation. (See Problems 5.10 to 5.12)

3. Integrals of the form $I_1 = \int_{-\infty}^{\infty} R(x) \sin x dx$, $I_2 = \int_{-\infty}^{\infty} R(x) \cos x dx$

   Here if $R(x) = P(x)/Q(x)$ has a zero of order two at infinity, and no poles on the axis, the previous method may be applied to the integral

$$\int_{-\infty}^{\infty} R(x)\, e^{ix} dx = I_3$$

   The real and imaginary parts of this integral determine the values of the integrals I_1 and I_2. Even when $R(x)$ has only a simple zero at infinity, the residue theorem may be used to evaluate I_3 (and thus I_1 and I_2). (See Problem 5.14)

4. Integrals of the form $\int_{-\infty}^{\infty} R(x)\, x^{\alpha} dx$, $(0 < \alpha < 1)$

Here we suppose the rational function has a zero of order two at infinity and at most a simple pole at the origin. This is sufficient to imply convergence of the improper integral. The complex contour integral in this case must take into consideration the branch cut of the multiple valued function z^{α}.

THE CAUCHY PRINCIPLE VALUE

The residue theorem can be used to evaluate various improper integrals including some integrals that are divergent in the ordinary sense. The improper integral of $f(x)$ on the interval $(-\infty, \infty)$ is said to be *convergent* if and only if f is integrable on $(-L, R)$ for every finite value of L and R and the integral

$$\int_{-L}^{R} f(x)\, dx$$

tends to a finite limit as L and R tend independently to infinity. We define the *Cauchy principle value* of the integral of $f(x)$ on the interval $(-\infty, \infty)$ to equal the limit

$$\lim_{R \to \infty} \int_{-R}^{R} f(x)\, dx$$

if the limit exists. Even when the improper integral is not convergent, the Cauchy principle value for the integral may exist. If the improper integral does converge then its value and the Cauchy principle value agree. If $f(x)$ is an even function of x then existence of the Cauchy principle value implies convergence of the improper integral. Note, however, that if $f(x)$ is an odd function then its Cauchy principle value is necessarily zero while the improper integral over $(-\infty, \infty)$ need not converge. When evaluating an improper integral by the residue theorem one should pay attention to whether the integral is actually convergent since the value obtained may be only the Cauchy principle value of a divergent integral.

THE ARGUMENT PRINCIPLE

We consider now another useful result involving integration around a simple closed contour. This result is known as the *argument principle*.

Theorem 5.3

Suppose $f(z)$ is analytic on and inside the simple closed contour C with the exception of pole type singularities lying inside C. Suppose further that $f(z)$ does not vanish on C but may have zeros inside C. If P and Z denote the number of poles and zeros of f lying inside C (each counted according to its multiplicity) then

$$\frac{1}{2\pi i}\int_C \frac{f'(z)}{f(z)}dz = \frac{1}{2\pi}\Delta \arg f(z) = Z - P \tag{5.7}$$

where $\Delta \arg f(z)$ denotes the change in the argument of $f(z)$ as z traces out the curve C in the positive sense.

The meaning of Theorem 5.3 is as follows. As z traces out the simple closed curve C in the z-plane, $w = f(z)$ traces out a corresponding closed curve Γ in the w-plane. If C enclosed Z zeros of $f(z)$ and no poles then Γ encircles the origin in the w-plane exactly Z times. Thus, the argument principle can be used to discover the location in the z-plane of the zeros of an analytic function. This can be very useful in determining stability of a linear system.

EXAMPLE 5.5

(a) Consider the function $f(z) = z^3 + 1$ with zeros $z_{1,2} = e^{i\pi/3}, e^{-i\pi/3}$ in the right half-plane and $z_3 = e^{i\pi}$ in the left half plane. Let the contour C consist of the segment, L, of the imaginary axis from $(0, iR)$ to $(0, -iR)$ together with the semicircular arc $C_R = \{z = Re^{i\vartheta}: -\pi/2 < \vartheta < \pi/2\}$ joining $(0, -iR)$ to $(0, iR)$. Since f is a polynomial it is entire and has no poles. In addition, for z on L, $f(iy) = 1 - iy^3$, $-R < y < R$, and it is clear that f has no zeros on L. Similarly, for $R > 0$ sufficiently large there are no zeros of f on C_R and thus f does not vanish on C. Then the argument principle applies. Although we already know that f has two zeros inside C we will apply the argument principle to illustrate how the computations are carried out.

On C_R we have $z = Re^{i\vartheta}$ and $w = R^3(e^{i3\vartheta} + R^{-3})$, hence $\varphi = \arg w$ tends to 3ϑ as R tends to infinity. Thus as ϑ increases from $-\pi/2$ to $\pi/2$ as z traces out C_R, φ increases from $-3\pi/2$ to $3\pi/2$; i.e., $\Delta\varphi = 3\pi/2 - (-3\pi/2) = 3\pi$ on C_R.

On L where $z = iy$ and $w = 1 - iy^3$ we have $\varphi = \arg w = \arctan(-y^3)$. Then at $z = (0, R)$, $\tan\varphi = -R^3$ is large and negative. Since φ must vary continuously as z traces out C, this indicates that we choose the branch of the arctan function where $\arctan(-R^3)$ tends to the value $3\pi/2$ as R tends to infinity. As z moves down the segment L from $(0, R)$ to $(0, -R)$, $\varphi = \arctan(-y^3)$ increases from $3\pi/2$ to $5\pi/2$, i.e., $\Delta\varphi = 5\pi/2 - 3\pi/2 = \pi$. Then the change in $\varphi = \arg f(z)$ as z traces out C is equal to $\Delta\varphi = \Delta\varphi(\text{on } C_R) + \Delta\varphi(\text{on } L) = 3\pi + \pi = 4\pi$. It follows then from (5.7) that there are $\Delta\varphi/2\pi = 2$ zeros of $f(z)$ inside C_R.

(b) Now let the contour C lie in the left half-plane where f has just one zero. We will see how the computions connected with the argument principle differ in order to indicate the presence of a single zero. Let C consist of the segment L, on the imaginary axis from $(0, -iR)$ to $(0, iR)$ together with the semicircular arc $C_R = \{z = Re^{i\vartheta}: -\pi/2 < \vartheta < \pi/2\}$ joining $(0, iR)$ to $(0, -iR)$. As z traces out C_R, ϑ increases from $\pi/2$ to $3\pi/2$ and φ increases from $3\pi/2$ to $9\pi/2$; i.e., $\Delta\varphi = 9\pi/2 - 3\pi/2 = 3\pi$ on C_R. As z moves up the imaginary axis from $(0, -R)$ to $(0, R)$, $\tan\varphi = -y^3$ decreases from a large positive number to a large negative number. In order that φ vary continuously, we choose the branch of arc $\tan\varphi$ so that arc $\tan\varphi$ decreases from $9\pi/2$ to $7\pi/2$; i.e., the change in φ on L is equal to $\Delta\varphi = 7\pi/2 - 9\pi/2 = -\pi$. Thus the change in $\varphi = \arg f(z)$ as z traces out C in this case is equal to $\Delta\varphi = \Delta\varphi(\text{on } C_R) + \Delta\varphi(\text{on } L) = 3\pi - \pi = 2\pi$. It follows then from (5.7) that there is $\Delta\varphi/2\pi = 1$ zero of $f(z)$ inside C_R.

ROUCHE'S THEOREM

A consequence of the argument principle, known as Rouche's theorem, is easier to use but is less powerful than the argument principle.

Theorem 5.4

Let $f(z)$ and $g(z)$ be analytic on and inside the simple closed contour C. If $|f(z)| < |g(z)|$ for z on C, then the functions $f(z)$ and $f(z) + g(z)$ have the same number of zeros (counting multiplicities) inside C.

EXAMPLE 5.6

Consider the function $P(z) = z^4 + 3z^3 + 1$. Letting $f(z) = 3z^3$ and $g(z) = z^4 + 1$, we see that $|f(z)| = 3$ on $C = \{|z| = 1\}$ and $|g(z)| \le |z^4| + 1 = 2$ for z on C. Since f and g are analytic on and inside C, it follows from Rouche's theorem that f and $P = f + g$ have the same number of zeros inside C. Since $f(z)$ has three zeros at the origin, $P(z)$ must have three zeros inside C. Note further that since $P(z)$ has real coefficients, any complex roots must occur in complex conjugate pairs. But P has three zeros inside C, and since $|z| = |z^*|$, at most two of these can be a complex conjugate pair. Hence at least two of the four zeros of $P(z)$ must be real zeros.

SOLVED PROBLEMS

Residues

PROBLEM 5.1

Suppose $f(z) = p(z)/q(z)$ where $p(z)$ and $q(z)$ are entire functions. If $q(z_0) = 0$ while $q'(z_0)$ and $p(z_0)$ are not zero then show that

$$\text{Res} f(z_0) = p(z_0)/q'(z_0) \tag{1}$$

SOLUTION 5.1

From (5.3) we have

$$\text{Res} f(z_0) = \lim_{z \to z_0} (z - z_0) f(z) = \lim_{z \to z_0} (z - z_0) \frac{p(z)}{q(z)}$$

$$= \lim_{z \to z_0} \frac{p(z)}{q(z)/(z - z_0)} = p(z_0)/q'(z_0)$$

PROBLEM 5.2

Suppose $f(z)$ has a pole of order m at z_0. Then show that $g(z) = (z - z_0)^m f(z)$ has a removable singularity at $z = z_0$ hence $g(z)$ can be defined at z_0 so as to be analytic in a neighborhood of z_0. Show further that

$$\text{Res} f(z_0) = g^{(m-1)}(z_0)/(m-1)! \tag{1}$$

SOLUTION 5.2

If $f(z)$ has a pole of order m at z_0 then

$$f(z) = \sum_{n=0}^{\infty} a_n (z - z_0)^n + \sum_{n=1}^{m} b_n (z - z_0)^{-n} \quad \text{and } b_m \neq 0.$$

It follows that

$$g(z) = (z - z_0)^m f(z) = b_m + b_{m-1}(z - z_0) + \ldots + \sum_{n=0}^{\infty} a_n (z - z_0)^{n+m}$$

has a removable singularity at z_0 and that $g(z)$ is analytic in a disc around z_0 if we define $g(z_0)$ to be b_m. Now, expanding $g(z)$ in a Taylor series about z_0

$$g(z) = g(z_0) + g'(z_o)(z - z_0) + \ldots + \frac{g^{(m-1)}(z_0)}{(m-1)!}(z - z_0)^{m-1} + \ldots \tag{2}$$

Thus, dividing (2) by $(z - z_0)^m$, we obtain the following expansion for f

$$f(z) = \frac{g(z)}{(z - z_0)^m} = \frac{g(z_o)}{(z - z_0)^m} + \ldots + \frac{g^{(m-1)}(z_0)}{(m-1)!}\frac{1}{(z - z_0)} + \ldots$$

The residue of $f(z)$ at z_0 is the coefficient of $(z - z_0)^{-1}$ in this expansion; i.e., the residue is given by (1).

PROBLEM 5.3

Find the residue of $f(z) = e^{iz}/(z^2 + a^2)$ at the singular point in the upper half plane if $a > 0$.

SOLUTION 5.3

Write

$$f(z) = \frac{e^{iz}}{(z - ia)(z + ia)} = \frac{e^{iz}/(z + ia)}{z - ia}$$

Then f has a simple pole at $z = ai$ in the upper half-plane. According to (5.4)

$$\operatorname{Res} f(ai) = e^{iz}/(z + ia)|_{z = ai} = e^{-a}/2ai.$$

PROBLEM 5.4

Find the residue of $f(z) = e^{iz}/(z^2 + 1)^2$ at the singular point in the lower half plane.

SOLUTION 5.4

If we write

$$f(z) = \frac{e^{iz}/(z - i)^2}{(z + i)^2}$$

then it is clear that f has a pole of order two at $z_0 = -i$ in the lower half plane. Then by the result of Problem 5.2

$$\operatorname{Res} f(-i) = \frac{d}{dz}(e^{iz}/(z - i)^2)\bigg|_{z = -i}$$

$$= \frac{ie^{iz}(z - i)^2 - e^{iz}2(z - i)}{(z - i)^4}\bigg|_{z = -i} = 0$$

PROBLEM 5.5

For $b > 0$, find the residue of $f(z) = e^{bz}/\cosh z$ at $z_0 = i\pi/2$.

SOLUTION 5.5

The poles of $f(z)$ occur at the zeros of $\cosh z$. These zeros are simple zeros and they are located at the points $z_n = (2n+1)\pi i/2$ for n an integer. Thus at $z_0 = i\pi/2$, we have by the result of Problem 5.1

$$\mathrm{Res}f(i\pi/2) = \frac{e^{ib\pi/2}}{\sinh(i\pi/2)} = \frac{e^{ib\pi/2}}{i\sin(\pi/2)} = -ie^{ib\pi/2}$$

The Residue Theorem

PROBLEM 5.6

Suppose that $f(z)$ is analytic on and inside a simple closed contour C except at isolated singular points $z_1, \ldots, z_m$, lying inside C. Then show that

$$\int_C f(z)\,dz = 2\pi i \sum_{j=1}^{m} \mathrm{Res}f(z_j) \tag{1}$$

SOLUTION 5.6

Since each singularity of f inside C is an isolated singularity, for each j, z_j may be enclosed in a circle $C_j = \{|z - z_j| = r_j\}$ lying inside C and having no points of overlap with any other circle C_k. Then $f(z)$ is analytic in the domain whose boundary is the positively oriented contour C together with the m negatively oriented circles $C_1, \ldots, C_m$. Then the Cauchy Goursat theorem implies

$$\int_C f(z)\,dz + \int_{-C_1} f(z)\,dz + \ldots + \int_{-C_m} f(z)\,dz = 0,$$

and

$$\int_C f(z)\,dz = \sum_{j=1}^{m} \int_{C_j} f(z)\,dz \tag{2}$$

Since $f(z)$ has a Laurent series expansion of the form (5.1) about each singular point z_j we have (see Problem 3.4)

$$\int_{C_j} f(z)\,dz = 2\pi i \mathrm{Res}f(z_j) \tag{3}$$

Substituting (3) into (2) yields (1).

PROBLEM 5.7

Use the residue theorem to evaluate the integral

$$\int_0^{2\pi} \frac{d\vartheta}{2 + \cos\vartheta}$$

SOLUTION 5.7

For z on the circle $C_j = \{|z| = 1\}$ we have $\cos\vartheta = (z + 1/z)/2$ and $dz = d\vartheta/iz$. Then

$$\int_0^{2\pi} \frac{d\vartheta}{2 + \cos\vartheta} = -i\int_C \frac{dz/z}{2 + (z + 1/z)/2} = -2i\int_C \frac{dz}{z^2 + 4z + 1}$$

Since

$$z^2 + 4z + 1 = (z + 2 - \sqrt{3})(z + 2 + \sqrt{3}) = (z - z_1)(z - z_2)$$

the integrand has simple poles at z_1 and z_2. Only the pole at z_1 lies inside C, however, and thus by the residue theorem

$$\int_C \frac{dz}{z^2 + 4z + 1} = 2\pi i \mathrm{Res} f(z_1) = 2\pi i\left(\frac{1}{2\sqrt{3}}\right) = \frac{\pi i}{\sqrt{3}}$$

Finally

$$\int_0^{2\pi} \frac{d\vartheta}{2 + \cos\vartheta} = 2\pi/\sqrt{3}$$

PROBLEM 5.8

Use the residue theorem to evaluate the integral

$$\int_0^{\pi} \frac{\sin^2\vartheta}{(5 - 4\cos\vartheta)} d\vartheta$$

SOLUTION 5.8

Since this integrand assumes the same values in the interval $(\pi, 2\pi)$ as it does in the interval $(0, \pi)$ it follows that

$$\int_0^{\pi} \frac{\sin^2\vartheta}{(5 - 4\cos\vartheta)} d\vartheta = \frac{1}{2}\int_0^{2\pi} \frac{\sin^2\vartheta}{(5 - 4\cos\vartheta)} d\vartheta = -\frac{i}{8}\int_C \frac{z^4 - 2z^2 + 1}{z^2(2z^2 - 5z + 2)} dz$$

where we have made the substitutions for $\sin\vartheta$ and $\cos\vartheta$ suggested by the previous problem. This integrand has simple poles at $z_1 = 1/2$ and $z_2 = 2$ and a pole of order two at $z_0 = 0$. Only z_0 and z_1 are inside $C = \{|z| = 1\}$. We compute

$$\mathrm{Res} f(z_0) = \frac{d}{dz} \frac{z^4 - 2z^2 + 1}{(2z^2 - 5z + 2)}\bigg|_{z=0} = 5/4$$

and

$$\mathrm{Res} f(z_1) = \frac{z^4 - 2z^2 + 1}{z^2(z - 2)}\bigg|_{z=1/2} = -3/4$$

Then

$$\int_0^{\pi} \frac{\sin^2\vartheta}{(5-4\cos\vartheta)}\,d\vartheta = -\frac{i}{8}2\pi i(5/4-3/4) = \pi/8$$

PROBLEM 5.9

Use the residue theorem to evaluate the convergent improper integral

$$\int_{-\infty}^{\infty} \frac{1}{(x^2+4)^2}\,dx \tag{1}$$

SOLUTION 5.9

Consider the complex contour integral

$$\int_{C_R} \frac{1}{(z^2+4)^2}\,dz \tag{2}$$

where C_R denotes the closed semicircular arc from $(-R, 0)$ to $(R, 0)$ along the real axis, and then along the semicircle $r = re^{i\vartheta}$, $0<\vartheta<\pi$. The integrand, $f(z) = (z^2+4)^2$ has poles of order two at $z_1 = 2i$ and $z_2 = -2i$, but for $R > 2$, only the pole at z_1 lies inside C_R. The residue of $f(z)$ at z_1 is equal to

$$\mathrm{Res}f(z_1) = \frac{d}{dz}(z+2i)^{-2}\Big|_{z=2i} = -2(4i)^{-3} = \frac{1}{32i}$$

and thus, by the residue theorem, the value of the contour integral (2) is $\pi/16$. To see that this is also the value of the integral (1) note that

$$\int_{C_R} \frac{1}{(z^2+4)^2}\,dz = \int_{-R}^{R} \frac{1}{(x^2+4)^2}\,dx + \int_0^{\pi} \frac{iRe^{i\vartheta}\,d\vartheta}{(R^2e^{i2\vartheta}+4)^2} \tag{3}$$

The value of the complex contour integral is equal to $\pi/16$ for all $R > 2$ by the deformation of contours result, Theorem 3.5. In particular, letting R tend to infinity, the first integral on the right in (3) tends to the integral (1). The second integral tends to the value zero with increasing R; i.e.,

$$\left|\int_0^{\pi} \frac{iRe^{i\vartheta}\,d\vartheta}{(R^2e^{i2\vartheta}+4)^2}\right| \le \frac{R}{(R^2-4)^2}\int_0^{\pi} d\vartheta \to 0 \text{ as } R \text{ tends to infinity.}$$

Note that the residue theorem proves that the Cauchy principle value exists and equals $\pi/16$. Since $f(x)$ is even, the integral of $f(x)$ from $-R$ to R is twice the integral from 0 to R, thus existence of the principle value implies convergence of the improper integral as well.

PROBLEM 5.10

Consider the complex contour integral

$$\int_{C_R} F(z)\,dz$$

where C_R denotes the closed semicircular arc from $(-R, 0)$ to $(R, 0)$ along the real axis, and then along the semicircle $r = Re^{i\vartheta}$, $0 < \vartheta < \pi$. Suppose for constants $M > 0$ and $p > 1$, $|F(z)| \le MR^{-p}$ for $z = Re^{i\vartheta}$. Then show that

$$\lim_{R \to \infty} \int_{C_R} F(z)\,dz = \int_{-\infty}^{\infty} F(x)\,dx \tag{1}$$

SOLUTION 5.10

Note that

$$\int_{C_R} F(z)\,dz - \int_{-R}^{R} F(x)\,dx = \int_0^{\pi} F(Re^{i\vartheta})\,iRe^{i\vartheta}\,d\vartheta$$

hence (1) follows from the fact that

$$\left|\int_0^{\pi} F(Re^{i\vartheta})\,iRe^{i\vartheta}\,d\vartheta\right| \le \int_0^{\pi} \left|F(Re^{i\vartheta})\,iRe^{i\vartheta}\right| |d\vartheta| \le \frac{M\pi}{R^{p-1}} \to 0 \text{ for } R \to \infty$$

Note that the hypotheses on $F(z)$ are sufficient to imply that the real integral on the right in (1) is a convergent improper integral.

PROBLEM 5.11

Use the residue theorem to compute the value of the improper integral

$$\int_{-\infty}^{\infty} \frac{1}{1+x^4}\,dx \tag{1}$$

SOLUTION 5.11

Note that for $z = Re^{i\vartheta}$,

$$\left|\frac{1}{1+z^4}\right| \le \frac{1}{R^4 - 1} = \frac{1}{R^4}\,\frac{1}{(1 - 1/R^4)} \le \frac{2}{R^4} \quad \text{for } R > 2$$

Then by the result of the previous problem, the improper integral is convergent and equals

$$\int_{C_R} \frac{1}{1+z^4}\,dz$$

for C_R the semicircular contour of the previous problem. Then $f(z)$ has simple zeros $z_m = \exp(i(2m+1)\pi/4)$, $m = 0, 1, 2, 3$ but only z_0 and

z_1 lie inside the contour. Problem 5.1 implies

$$\text{Res} f(z_0) = \frac{1}{4(z_0)^3} = \frac{1}{4} e^{-i3\pi/4} \quad \text{Res} f(z_1) = \frac{1}{4} e^{-i9\pi/4} = \frac{1}{4} e^{-i\pi/4}$$

and thus, by the residue theorem,

$$\int_{-\infty}^{\infty} \frac{1}{1+x^4} dx = \int_{C_R} \frac{1}{1+z^4} dz$$

$$= \frac{2\pi i}{4} (\cos\pi/4 - i\sin\pi/4 + \cos 3\pi/4 - i\sin 3\pi/4)$$

$$= \pi \sin\pi/4 = \pi\sqrt{2}/2$$

PROBLEM 5.12

Use the residue theorem to compute the value of the improper integral

$$\int_{-\infty}^{\infty} \frac{x^2}{(x^2+1)(x^2+2x+2)} dx \tag{1}$$

SOLUTION 5.12

The integrand $f(x)$ in (1) is a rational function whose denominator has degree two more than the degree of the numerator. Thus the conditions of Problem 5.10 are satisfied and the real integral (1) is has value equal to the contour integral of $f(z)$ around the semicircular contour C_R. The integrand $f(z)$ has simple poles at $z_0 = i$ and $z_1 = -1+i$. Then

$$\text{Res} f(z_0) = \left. \frac{z^2}{(z+1)(z^2+2z+2)} \right|_{z=i} = \frac{2+i}{10}$$

$$\text{Res} f(z_1) = \left. \frac{z^2}{(z+1+i)(z^2+1)} \right|_{z=-1+i} = \frac{-1-2i}{5}$$

and the value of the integral in (1) is equal to $2\pi i(-3i/10) = 3\pi/5$.

PROBLEM 5.13

Use the residue theorem to compute the value of the improper integral

$$\int_{-\infty}^{\infty} \frac{\cos x}{(x^2+1)^2} dx \tag{1}$$

SOLUTION 5.13

The integral (1) is the real part of the integral

$$\int_{-\infty}^{\infty} \frac{e^{ix}}{(x^2+1)^2} dx \tag{2}$$

which can be evaluated by integrating the complex valued function $f(z) = e^{iz}(z^2+1)^{-2}$ over the semicircular contour C_R used in previous problems. Only the pole of order two $z_0 = i$ lies inside the contour and thus the value of the integral is given by

$$\int_{C_R} e^{iz}(z^2+1)^{-2} = 2\pi i \text{Res} f(i) = 2\pi i \frac{d}{dz}\left(e^{iz}(z^2+1)^{-2}\right)\Big|_{z=i} = \frac{\pi}{e}$$

$$\int_{C_R} e^{iz}(z^2+1)^{-2} = \int_{-R}^{R} \frac{e^{ix}}{(x^2+1)^2} dx + \int_0^{\pi} \frac{e^{iRe^{-\vartheta}}}{(R^2 e^{2i\vartheta}+1)^2} iRe^{i\vartheta} d\vartheta \tag{3}$$

and we can show (see Problem 5.14) that

$$\left|\int_0^{\pi} \frac{e^{iRe^{-\vartheta}}}{(R^2 e^{2i\vartheta}+1)^2} iRe^{i\vartheta} d\vartheta\right| \to 0 \text{ as } R \text{ tends to infinity}$$

Then as R tends to infinity the contour integral of $f(z)$ over C_R tends to the integral (2). But for all values of R, the contour integral has value π/e and it follows that the integral (2) equals π/e. Finally, since $\sin x\,(x^2+1)^{-2}$ is an odd function, we find

$$\int_{-\infty}^{\infty} \frac{e^{ix}}{(x^2+1)^2} dx = \int_{-\infty}^{\infty} \frac{\cos x}{(x^2+1)^2} dx + i\int_{-\infty}^{\infty} \frac{\sin x}{(x^2+1)^2} dx$$

$$= \int_{-\infty}^{\infty} \frac{\cos x}{(x^2+1)^2} dx + 0 = \pi/e.$$

PROBLEM 5.14

Consider the complex contour integral

$$\int_{C_R} F(z)\, e^{ibz} dz \quad b > 0,$$

where C_R denotes the closed semicircular arc from $(-R, 0)$ to $(R, 0)$ along the real axis, and then along the semicircle $r = Re^{i\vartheta}$, $0 < \vartheta < \pi$. Suppose for constants $M > 0$ and $p > 0$, $|F(z)| \le MR^{-p}$ for $z = Re^{i\vartheta}$. Then show that for any positive constant b,

$$\lim_{R \to \infty} \int_{C_R} F(z)\, e^{ibz} dz = \int_{-\infty}^{\infty} F(x)\, e^{ibx} dx \tag{1}$$

SOLUTION 5.14

Note that

$$\int_{C_R} F(z)\, e^{ibz} dz - \int_{-R}^{R} F(x)\, e^{ibx} dx = \int_0^{\pi} F(Re^{i\vartheta})\, e^{ibRe^{i\vartheta}}\, iRe^{i\vartheta} d\vartheta \tag{2}$$

hence (1) follows from the fact that the integral on the right in (2) tends to zero as R tends to infinity. To see this write

$$\left| \int_0^{\pi} F(Re^{i\vartheta})\, e^{ibRe^{i\vartheta}}\, iRe^{i\vartheta} d\vartheta \right| \le \int_0^{\pi} \left| F(Re^{i\vartheta}) \right| Re^{-bR\sin\vartheta} d\vartheta$$

$$\le \frac{2M}{R^{p-1}} \int_0^{\pi/2} Re^{-bR\sin\vartheta} d\vartheta$$

Now we use the fact that $2/\pi \le (\sin\vartheta)/\vartheta \le 1$ for $0 \le \vartheta \le \pi/2$ to conclude

$$\left| \int_0^{\pi} F(Re^{i\vartheta})\, e^{ibRe^{i\vartheta}}\, iRe^{i\vartheta} d\vartheta \right| \le \frac{M\pi}{bR^{p}} (1 - e^{-bR}) \to 0 \text{ as } R \to \infty$$

Note that this argument proves that the integral

$$\int_{-R}^{R} F(x)\, e^{ibx} dx \tag{3}$$

tends to a finite limit as R tends to infinity. We refer to this limiting value as the *Cauchy principle value* of the integral. Showing that the Cauchy principle value limit exists for an integral is not the same as showing that the improper integral converges. To show that we must show

$$\int_{-L}^{R} F(x)\, e^{ibx} dx \tag{4}$$

tends to a finite limit as L and R tend (independently) to infinity. The given hypotheses on $F(z)$ are sufficient to imply the existence of the limit in (3) as R tends to infinity and by a more complicated argument, they imply also the existence of the limit of (4) as R and L tend to infinity; i.e., they imply the convergence of the improper integral.

PROBLEM 5.15

Use the residue theorem to evaluate the improper integral

$$\int_0^{\infty} \frac{x \sin x}{x^2 + 1} dx \tag{1}$$

SOLUTION 5.15

For z on the real axis, the real part of the complex valued function $f(z) = ze^{iz}(z^2+1)^{-1}$ reduces to the integrand in (1). Thus for C_R the semicircular contour used in previous problems

$$\int_{C_R} \frac{ze^{iz}}{z^2+1}dz = \int_{-R}^{R} \frac{xe^{ix}}{x^2+1}dx + \int_0^{\pi} \frac{Re^{i\vartheta + iRe^{i\vartheta}}}{R^2e^{i2\vartheta}+1} iRe^{i\vartheta}d\vartheta$$

Note that the result in Problem 5.14 applies to this integrand $f(z)$ and thus we have

$$\lim_{R\to\infty}\int_{C_R} \frac{ze^{iz}}{z^2+1}dz = \int_{-\infty}^{\infty} \frac{xe^{ix}}{x^2+1}dx$$

$$= \int_{-\infty}^{\infty} \frac{x\cos x}{x^2+1}dx + i\int_{-\infty}^{\infty} \frac{x\sin x}{x^2+1}dx$$

$$= 0 + 2i\int_0^{\infty} \frac{x\sin x}{x^2+1}dx$$

Here we have used the fact that $x\cos x\,(x^2+1)^{-1}$ is an odd function while the integrand $x\sin x\,(x^2+1)^{-1}$ is even. Only the simple pole for $f(z)$ located at $z = i$ is inside C_R and since

$$\mathrm{Res}f(i) = \left.\frac{ze^{iz}}{z+i}\right|_{z=i} = e^{-1}/2$$

we have

$$\int_{C_R} \frac{ze^{iz}}{z^2+1}dz = 2\pi i\frac{1}{2e} = 2i\int_0^{\infty} \frac{x\sin x}{x^2+1}dx$$

Then it is clear that the integral in (1) has the value $\pi/2e$.

PROBLEM 5.16

Suppose $f(z)$ has a simple pole at $z = z_0$ and for $\varepsilon > 0$, $\vartheta_1 < \vartheta_2$, let C_ε denote the circular arc $\{z\colon |z-z_0| = \varepsilon,\ \vartheta_1 < \arg(z-a_0) < \vartheta_2\}$ shown in Figure 5.1.

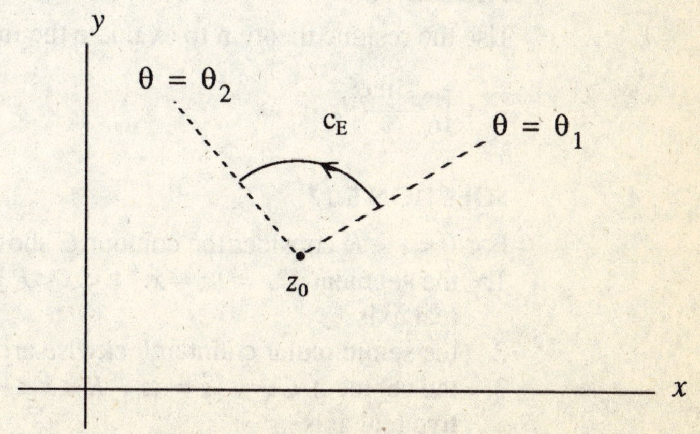

Figure 5.1

Then show that

$$\lim_{\varepsilon \to 0} \int_{C_\varepsilon} f(z)\, dz = (\vartheta_2 - \vartheta_1)\, i \operatorname{Res} f(z_0) \tag{1}$$

Note that when $\vartheta_2 - \vartheta_1 = 2\pi$ the arc C_ε is a full circle of radius ε and (1) reduces to the usual expression of the residue theorem.

SOLUTION 5.16

If $f(z)$ has a simple pole at z_0 then f has a Laurent series expansion of the form

$$f(z) = \frac{b_1}{z - z_0} + \sum_{n=0}^{\infty} a_n (z - z_0)^n \tag{2}$$

converging in a deleted neighborhood of z_0. Then for $\varepsilon > 0$ sufficiently small,

$$\int_{C_\varepsilon} f(z)\, dz = b_1 \int_{C_\varepsilon} \frac{dz}{z - z_0} + \sum_{n=0}^{\infty} a_n \int_{C_\varepsilon} (z - z_0)^n dz$$

$$= b_1 \int_{\vartheta_1}^{\vartheta_2} \frac{1}{\varepsilon e^{i\vartheta}} i\varepsilon e^{i\vartheta} d\vartheta + \sum_{n=0}^{\infty} a_n \int_{\vartheta_1}^{\vartheta_2} \varepsilon^{n+1} e^{i(n+1)\vartheta} i d\vartheta$$

$$= ib_1 (\vartheta_2 - \vartheta_1) + \sum_{n=0}^{\infty} \frac{a_n \varepsilon^{n+1}}{n+1} \left(e^{i(n+1)\vartheta_2} - e^{i(n+1)\vartheta_1}\right)$$

and

$$\int_{C_\varepsilon} f(z)\, dz \to ib_1 (\vartheta_2 - \vartheta_1) \quad \text{as } \varepsilon \to 0.$$

Since $b_1 = \operatorname{Res} f(z_0)$ this proves (1).

PROBLEM 5.17

Use the residue theorem to evaluate the improper integral

$$\int_0^{\infty} \frac{\sin x}{x} dx \tag{1}$$

SOLUTION 5.17

For $0 < \varepsilon < R$ consider the contour C shown in Figure 5.2, consisting of:

1. the segment $C_1 = \{z = x\colon\ \varepsilon < x < R\}$ from ε to R along the positive real axis
2. the semicircular counterclockwise arc $C_R = \{z = Re^{i\vartheta}\colon\ 0 < \vartheta < \pi\}$
3. the segment $C_2 = \{z = x\colon\ -R < x < -\varepsilon\}$ from $-R$ to $-\varepsilon$ on the negative real axis
4. the semicircular clockwise are $C_\varepsilon = \{z = \varepsilon e^{i\vartheta}\colon\ \pi > \vartheta > 0\}$

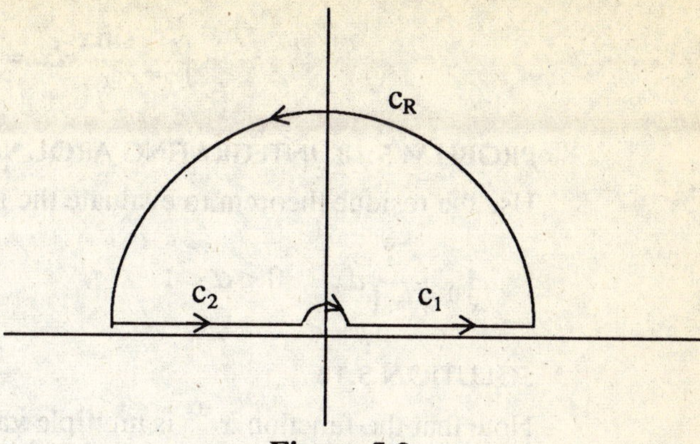

Figure 5.2

Then the Cauchy Goursat theorem implies that for all ε, R such that $0 < \varepsilon < R$

$$\int_C \frac{e^{iz}}{z} dz = 0 \tag{2}$$

since the simple pole at $z = 0$ lies outside C. But

$$\int_C \frac{e^{iz}}{z} dz = \int_{C_1} \frac{e^{iz}}{z} dz + \int_{C_R} \frac{e^{iz}}{z} dz + \int_{C_2} \frac{e^{iz}}{z} dz + \int_{C_\varepsilon} \frac{e^{iz}}{z} dz \tag{3}$$

and

$$\int_{C_R} \frac{e^{iz}}{z} dz \to 0 \text{ as } R \to \infty \text{ by Problem 5.14} \tag{4}$$

$$\int_{-C_\varepsilon} \frac{e^{iz}}{z} dz \to \pi i\, (e^{iz}|_{z=0}) = \pi i \text{ as } \varepsilon \to 0 \text{ by Problem 5.16} \tag{5}$$

In addition,

$$\int_{C_1} \frac{e^{iz}}{z} dz + \int_{C_2} \frac{e^{iz}}{z} dz = \int_\varepsilon^R \frac{e^{ix}}{x} dx + \int_{-R}^{-\varepsilon} \frac{e^{ix}}{x} dx$$

$$= 2i \int_\varepsilon^R \frac{\sin x}{x} dx \tag{6}$$

where we used the fact that $x^{-1}\sin x$ and $x^{-1}\cos x$ are even and odd functions, respectively. Since (6) holds for all $R > \varepsilon > 0$, we can let ε tend to zero and R tend to infinity. Then using (4), (5) and (6) in (3), leads to

$$2i \int_{-\infty}^{\infty} \frac{\sin x}{x} dx - \pi i + 0 = 0;$$

i.e.,

$$\int_{-\infty}^{\infty} \frac{\sin x}{x}\, dx = \pi/2.$$

PROBLEM 5.18 INTEGRATING AROUND A BRANCH CUT

Use the residue theorem to evaluate the improper integral

$$\int_0^{\infty} \frac{x^{-\alpha}}{x+1}\, dx \qquad 0<\alpha<1 \tag{1}$$

SOLUTION 5.18

Note that the function $x^{-\alpha}$ is multiple valued. We choose a branch of this function by placing a branch cut along the positive real axis and then we consider the contour integral

$$\int_C \frac{z^{-\alpha}}{z+1}\, dz \tag{2}$$

where C denotes the contour shown in Figure 5.3. That is C is composed of:

1. the segment $C_1 = \{z = x\text{: } \varepsilon < x < R\}$ from ε to R along the positive real axis
2. the circular counterclockwise arc $C_R = \{z = Re^{i\vartheta}\text{: } 0 < \vartheta < 2\pi\}$
3. the segment $C_2 = \{z = x\text{: } R > x > \varepsilon\}$ from R to ε on the positive real axis
4. the circular clockwise arc $C_\varepsilon = \{z = \varepsilon e^{i\vartheta}\text{: } 2\pi > \vartheta > 0\}$

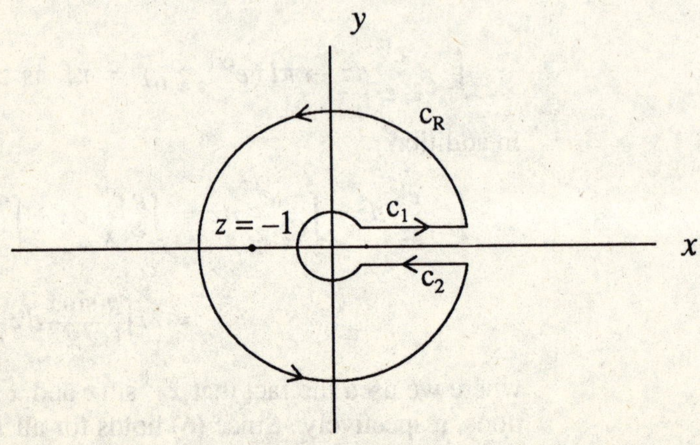

Figure 5.3

Thus the contour avoids the branch cut and the branch point at the origin but includes the simple pole at $z = -1$ in its interior. For all $R > \varepsilon > 0$, the value of the complex contour integral in (2) equals

$$2\pi i \operatorname{Res} f(-1) = 2\pi i z^{-\alpha}\Big|_{z=-1} = 2\pi i \left(e^{i\pi}\right)^{-\alpha} \tag{2}$$

But we also have

$$\int_C \frac{z^{-\alpha}}{z+1}\,dz = \int_{C_1} + \int_{C_R} + \int_{C_2} + \int_{C_\varepsilon} \tag{3}$$

and

$$\int_{C_1} \frac{z^{-\alpha}}{z+1}\,dz = \int_\varepsilon^R \frac{(xe^0)^{-\alpha}}{x+1}\,dx = \int_\varepsilon^R \frac{x^{-\alpha}}{x+1}\,dx \tag{4}$$

$$\int_{C_R} \frac{z^{-\alpha}}{z+1}\,dz = \int_0^{2\pi} \frac{R^{-\alpha}e^{-i\alpha\vartheta}}{Re^{i\vartheta}+1}\, iRe^{i\vartheta}\,d\vartheta \tag{5}$$

$$\int_{C_2} \frac{z^{-\alpha}}{z+1}\,dz = \int_R^\varepsilon \frac{(xe^{2\pi i})^{-\alpha}}{x+1}\,dx = -e^{i2\pi\alpha}\int_\varepsilon^R \frac{x^{-\alpha}}{x+1}\,dx \tag{6}$$

$$\int_{C_\varepsilon} \frac{z^{-\alpha}}{z+1}\,dz = \int_{2\pi}^{0} \frac{\varepsilon^{-\alpha}e^{-i\alpha\vartheta}}{\varepsilon e^{i\vartheta}+1}\, i\varepsilon e^{i\vartheta}\,d\vartheta \tag{7}$$

But

$$\left|\int_0^{2\pi} \frac{R^{-\alpha}e^{-i\alpha\vartheta}}{Re^{i\vartheta}+1}\, iRe^{i\vartheta}\,d\vartheta\right| \le 2R^{-\alpha}2\pi \quad \text{for } R>2 \tag{8}$$

and

$$\left|\int_{2\pi}^{0} \frac{\varepsilon^{-\alpha}e^{-i\alpha\vartheta}}{\varepsilon e^{i\vartheta}+1}\, i\varepsilon e^{i\vartheta}\,d\vartheta\right| \le 2\varepsilon^{1-\alpha}2\pi \quad \text{for } 0<\varepsilon<1/2 \tag{9}$$

Thus for $0<\alpha<1$, the integrals over C_R and C_ε tend to zero as R tends to infinity and as ε tends to zero. Then (3) reduces to

$$\int_C \frac{z^{-\alpha}}{z+1}\,dz = 2\pi i e^{-i\pi\alpha} = \left(1-e^{i2\pi\alpha}\right)\int_0^\infty \frac{x^{-\alpha}}{x+1}\,dx$$

Then

$$\int_0^\infty \frac{x^{-\alpha}}{x+1}\,dx = \frac{2\pi i e^{-i\pi\alpha}}{\left(1-e^{i2\pi\alpha}\right)} = \frac{\pi}{\sin \pi\alpha}$$

PROBLEM 5.19

Use the residue theorem to show that

$$\int_0^\infty \sin x^2 dx = \int_0^\infty \cos x^2 dx = \frac{1}{2}\sqrt{\pi/2} \tag{1}$$

SOLUTION 5.19

Consider the integral

$$\int_C e^{iz^2} dz = \int_{C_1} + \int_{C_R} + \int_{C_2} \tag{2}$$

where C denotes the wedge shaped contour shown in Figure 5.4. Her C consists of:

1. the segment C_1 from 0 to R along the positive real axis
2. the circular counterclockwise arc $C_R = \{z = Re^{i\vartheta}: 0 < \vartheta < \pi/4\}$
3. the segment C_2 from $r = R$ to $r = 0$ along the ray $\vartheta = \pi/4$

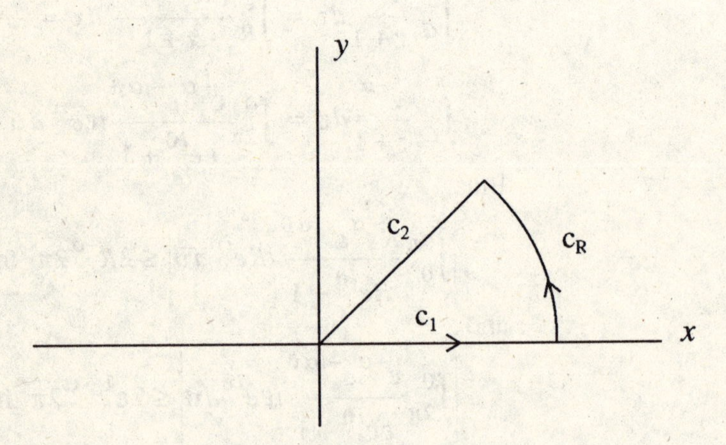

Figure 5.4

Since e^{iz^2} is entire, the contour integral in (2) equals zero by the Cauchy Goursat theorem. In addition, by an argument similar to the one used in Problem 5.14, we can show that the integral over C_R tends to zero as R tends to infinity. Then (2) implies that as R tends to infinity, we obtain in the limit

$$\int_C e^{iz^2} dz = \int_0^\infty e^{ix^2} dx + \int_0^\infty e^{ir^2 e^{i\pi/2}} e^{i\pi/4} dr = 0. \tag{3}$$

But

$$\int_0^\infty e^{ix^2} dx = \int_0^\infty (\cos x^2 + i \sin x^2)\, dx \tag{4}$$

and

$$\int_{\infty}^{0} e^{ir^2 e^{i\pi/2}} e^{i\pi/4} dr = -e^{i\pi/4} \int_{0}^{\infty} e^{-r^2} dr = -(\cos\frac{\pi}{4} + i\sin\frac{\pi}{4})\sqrt{\pi/2} \tag{5}$$

Substituting (4) and (5) in (3) and equating real and imaginary parts of the resulting equation leads to (1). Note that we have used

$$\int_{0}^{\infty} e^{-r^2} dr = \sqrt{\pi/2}.$$

This definite integral can be evaluated by elementary means.

PROBLEM 5.20

Use the residue theorem to evaluate the improper integral

$$\int_{0}^{\infty} \frac{\cosh px}{\cosh x} dx, \quad 0 < p < 1 \tag{1}$$

SOLUTION 5.20

Consider the complex contour integral

$$\int_{C} \frac{e^{pz}}{\cosh z} dz$$

where C denotes the rectangular contour shown in Figure 5.5. The poles of $f(z) = e^{pz}/\cosh z$ are simple poles at the points $z_n = (2n+1)i\pi/2$ for integer values of n (see Problem 5.5). Only the point z_0 lies inside C and by the result of Problem 5.5,

$$\text{Res} f(z_o) = \frac{e^{pz_0}}{\sinh z_0} = -ie^{ip\pi/2} \tag{2}$$

Thus

$$\int_{C} \frac{e^{pz}}{\cosh z} dz = 2\pi e^{ip\pi/2} \tag{3}$$

In addition, referring to the definition of the contour C,

$$\int_{C} \frac{e^{pz}}{\cosh z} dz = \int_{-R}^{R} \frac{e^{px}}{\cosh x} dx + \int_{0}^{\pi} \frac{e^{p(R+iy)}}{\cosh(R+iy)} i dy$$
$$+ \int_{R}^{-R} \frac{e^{p(x+i\pi)}}{\cosh(x+i\pi)} dx + \int_{\pi}^{-\pi} \frac{e^{p(-R+iy)}}{\cosh(-R+iy)} i dy$$

But

$$\left| \int_{0}^{\pi} \frac{e^{p(R+iy)}}{\cosh(R+iy)} i dy \right| \le \frac{2e^{pR}}{e^{R} - e^{-R}} \pi < 4\pi e^{(p-1)R} \quad \text{for } R > \frac{1}{2}\ln 2$$

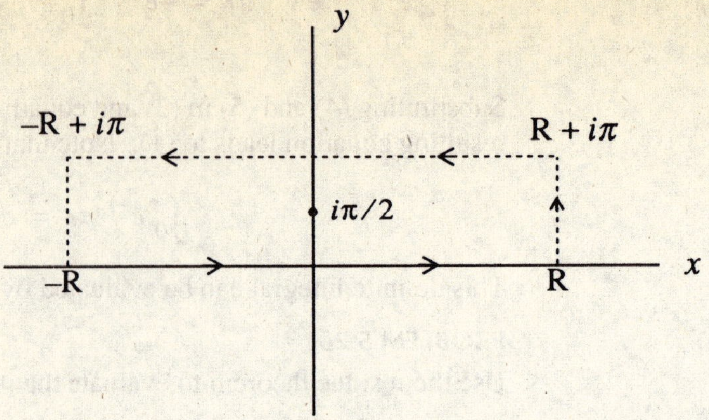

Figure 5.5

and for $0 < p < 1$ this integral clearly tends to zero as R tends to infinity. By a similar argument, the integral over the other vertical side of the rectangular contour tends to zero with increasing R. Since the result (3) holds for all $R > 0$, we have

$$\int_{-\infty}^{\infty} \frac{e^{px}}{\cosh x} dx + \int_{\infty}^{-\infty} \frac{e^{p(x+i\pi)}}{\cosh(x+i\pi)} dx = 2\pi e^{ip\pi/2};$$

i.e., since $\cosh(x+i\pi) = -\cosh x$,

$$\int_{-\infty}^{\infty} \frac{e^{px}}{\cosh x} dx = \frac{2\pi e^{ip\pi/2}}{1+e^{ip\pi}} = \frac{\pi}{\cos(p\pi/2)}$$

Finally, note that

$$\int_{-\infty}^{\infty} \frac{e^{px}}{\cosh x} dx = \int_{-\infty}^{0} \frac{e^{px}}{\cosh x} dx + \int_{0}^{\infty} \frac{e^{px}}{\cosh x} dx$$

$$= \int_{0}^{\infty} \frac{e^{-px}}{\cosh x} dx + \int_{0}^{\infty} \frac{e^{px}}{\cosh x} dx$$

$$= 2\int_{0}^{\infty} \frac{\cosh px}{\cosh x} dx$$

Thus

$$\int_{0}^{\infty} \frac{\cosh px}{\cosh x} dx = \frac{\pi/2}{\cos(p\pi/2)}$$

Zeros of Polynomials

PROBLEM 5.21

Sketch the curve traced out in the w-plane by $w = z^3 + 1$ as z traces out each of the following circles in the z-plane:

(a) $|z-(-1-i)| = 1/2$ (contains no zeros of $f(z) = z^3+1$)

(b) $|z+1| = 1/2$ (contains one zero of $f(z) = z^3+1$)

(c) $|z-1/2| = 1$ (contains two zeros of f)

(d) $|z-(0.1+0.1i)| = 2$ (contains three zeros of f)

SOLUTION 5.21

The curves in the w-plane are shown in Figure 5.6 where it can be seen that the curve circles the origin no times in the case of Figure 5.6(a) when no zeros of $f(z) = z^3+1$ are inside the corresponding circle in the z-plane. In the case that the z-plane circle includes one, two or three zeros of $f(z) = z^3+1$, the curve can be seen to circle the origin one, two or three times. This is a visual demonstration of the argument principle.

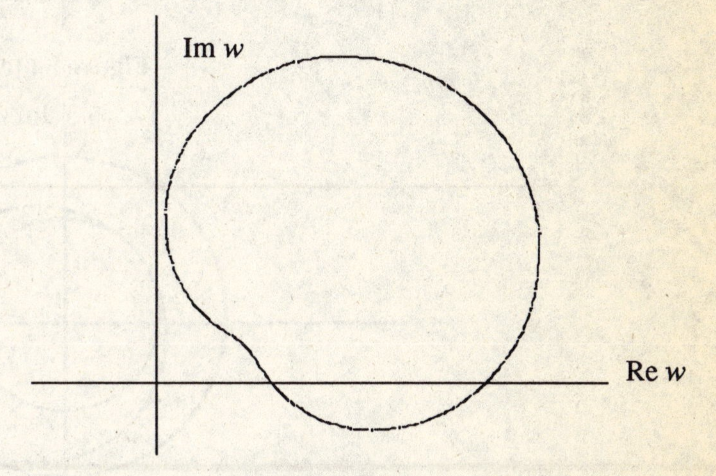

Figure 5.6(a)

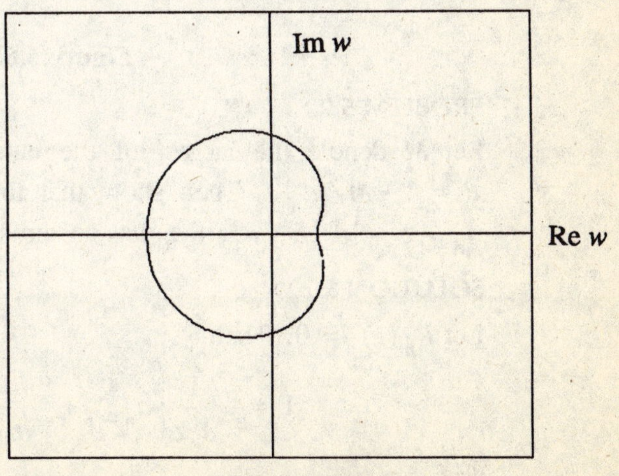

Figure 5.6(b)

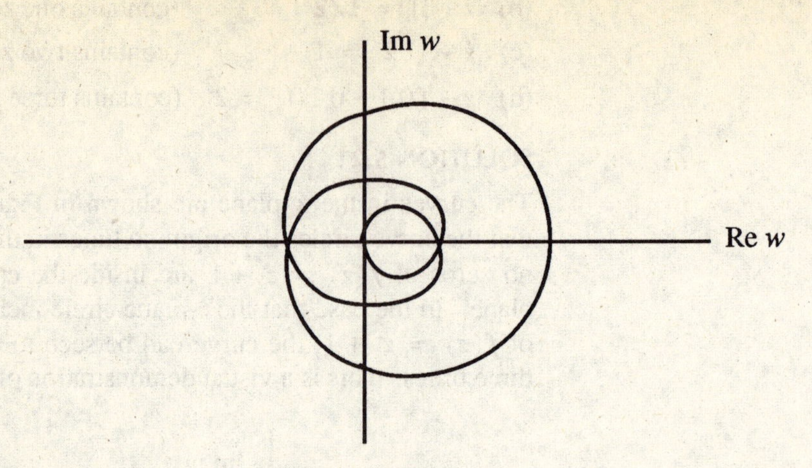

Figure 5.6(c)

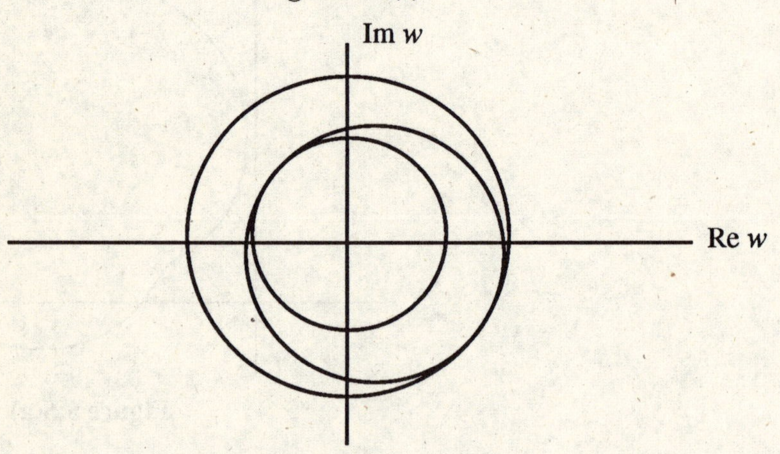

Figure 5.6(d)

PROBLEM 5.22

Let M denote the largest of the numbers $|a_n|, \ldots, |a_1|, |a_0|$ and let $R = 1 + M/|a_n|$. Then show that the polynomial $P_n(z) = a_n z^n + a_{n-1} z^{n-1} + \ldots + a_1 z + a_0$ has no zeros for which $|z| \geq R$.

SOLUTION 5.22

Let $P_n(z) = 0$. Then

$$1 = -\frac{1}{a_n z}\left(a_{n-1} + \frac{a_{n-2}}{z} + \ldots + \frac{a_0}{z^{n-1}}\right)$$

and

$$1 \le \frac{1}{|a_n||z|}\left(|a_{n-1}| + \frac{|a_{n-2}|}{|z|} + \ldots + \frac{|a_0|}{|z|^{n-1}}\right)$$

$$\le \frac{M}{|a_n||z|}\left(1 + \frac{1}{z} + \ldots + \frac{1}{|z|^{n-1}}\right)$$

$$< \frac{M}{|a_n||z|}\sum_{m=0}^{\infty} |z|^{-m} = \frac{M}{|a_n|}\frac{1}{|z|-1}$$

From this last inequality we conclude $|z| < 1 + M/|a_n| = R$. Thus all of the zeros of $P_n(z)$ are contained in a disc of radius R where R depends on the coefficients.

PROBLEM 5.23

Consider the function $P(z) = z^3 + z^2 + 4z + 1$ and determine how many zeros of P are located in the right half plane or on the imaginary axis.

SOLUTION 5.23

By the result of the previous problem, the zeros of P are all contained in a disc or radius 5. In addition, on the imaginary axis we have $P(iy) = 1 - y^2 + i(4y - y^3)$ and clearly there is no value of y for which the real and imaginary parts of $P(iy)$ vanish simultaneously.

Let the contour C consist of the segment, L, of the imaginary axis from $(0, iR)$ to $(0, -iR)$ together with the semicircular arc $C_R = \{z = Re^{i\vartheta}: -\pi/2 < \vartheta < \pi/2\}$ joining $(0, -iR)$ to $(0, iR)$. There are no zeros of $P(z)$ on L and, for $R > 5$, there are no zeros of P on C_R. Note that for z on C_R. Note that for z on C_R

$$w = P(z) = R^3 e^{i3\vartheta} + R^2 e^{i2\vartheta} + 4Re^{i\vartheta} + 1$$

$$= R^3 (e^{i3\vartheta} + R^{-1} e^{i2\vartheta} + 4R^{-2} e^{i\vartheta} + R^{-3})$$

Thus for large R, w is approximately equal to $R^3 e^{i3\vartheta}$. Then as ϑ increases from $-\pi/2$ to $\pi/2$ when z traverses C_R, $\varphi = \arg w$ increases from $-3\pi/2$ to $3\pi/2$; i.e., $\Delta\varphi(\text{on } C_R) = 3\pi/2 - (-3\pi/2) = 3\pi$. In general, if $P(z)$ is a polynomial of degree n, then $\Delta\varphi(\text{on } C_R) = n\pi$.

On the part L of the contour we have $P(iy) = 1 - y^2 + iy(4 - y^2)$ which implies

$$\varphi = \arg w = \text{arc tan}\frac{y(4-y^2)}{1-y^2}\text{; i.e., } \tan\varphi = \frac{y(4-y^2)}{1-y^2}$$

Then we begin with $\varphi = 3/2$ when y is large and positive. As y decreases to the value 4, $\tan\varphi$ decreases from a large positive value to 0. Correspondingly, φ decreases from $3\pi/2$ to the value π. As y decreases still further to the value 1, $\tan\varphi$ tends to minus infinity and φ decreases to $\pi/2$. As y decreases further still from 1 to 0, $\tan\varphi$ switches sign and we move onto the next branch of the tan function. Thus as y decreases from 1 to 0, $\tan\varphi$ decreases from infinity to zero and φ foes from $\pi/2$ down to 0. Then as z moves down L from iR to 0, φ decreases from $3\pi/2$ to 0.

Now $P(z)$ has real coefficients so $P(z^*) = P(z)^*$ and $P(-iy) = -P(iy)$. Then the trace of $P(iy)$ for $0 < y < R$ may be reflected through the real axis to get the trace of $P(iy)$ for $0 > y > -R$. In particular, as z moves along L from 0 to $-iR$, φ must decrease from 0 to $-3\pi/2$. Then $\Delta\varphi(\text{on } L) = -3\pi/2 - 3\pi/2 = -3\pi$.

Then the change in $\varphi = \arg P(z)$ as z traces out C is equal to

$$\Delta\varphi = \Delta\varphi(\text{on } C_R) + \Delta\varphi(\text{on } L) = 3\pi - 3\pi = 0.$$

Then, by the argument principle, $P(z)$ has no zeros in the right half plane.

PROBLEM 5.24

Consider the function $P(z) = z^3 - z^2 + z + 4$ and determine how many zeros of P are located in the right half plane or on the imaginary axis.

SOLUTION 5.24

As in the previous problem all the zeros of $P(z)$ are contained in a disc of radius 5 and there are no zeros on the imaginary axis. In addition, since the degree of $P(z)$ is 3, we have $\Delta\varphi(\text{on } C_R) = 3\pi$ for the semicircular arc C_R.

As z moves along L from $(0, iR)$ to $(0, -iR)$ we have

$$\varphi = \arg w = \arctan\frac{y(1-y^2)}{4+y^2}\text{; i.e., } \tan\varphi = \frac{y(1-y^2)}{4+y^2} \qquad (1)$$

Then we begin with φ decreasing from the value $3\pi/2$ when y is large and positive. Then we see from (1) that $\tan\varphi$ is negative and as y decreases to the value 1, $\tan\varphi$ increases to zero and φ increases from $3\pi/2$ to 2π. As y decreases further still from 1 to 0, $\tan\varphi$ becomes positive and then returns to zero. Thus φ returns to the value 2π at $y = 0$, and as z moves down L from iR to 0, φ increases from $3\pi/2$ to 2π.

Similarly, as y decreases from 0 to -1, $\tan\varphi$ goes negative and then returns to zero and $\varphi = 2\pi$. As y tends to negative infinity, $\tan\varphi$ increases steadily to plus infinity and φ tends to the value $5\pi/2$. Thus

$$\Delta\varphi(\text{on } L) = 5\pi/2 - 3\pi/2 = \pi$$

and

$$\Delta\varphi = \Delta\varphi(\text{on } C_R) + \Delta\varphi(\text{on } L) = 3\pi + \pi = 4\pi.$$

Then the argument principle implies there are two zeros of $P(z)$ in the right half plane.

PROBLEM 5.25

Prove Theorem 5.3, the argument principle.

SOLUTION 5.25

Let $F(z) = f'(z)/f(z)$. Then under the hypotheses of Theorem 5.3 the only singular points of F occur at the singular points and zeros of f. Then the residue theorem implies

$$\int_C F(z)\,dz = 2\pi i \text{ (sum of residues at poles and zeros of } f \text{ inside } C\text{)}$$

Now if z_0 is a zero of order m for $f(z)$, then in a neighborhood of z_0 we have $f(z) = (z-z_0)^m g(z)$ for $g(z)$ analytic at z_0 and $g(z_0)$ different from zero. Then

$$F(z) = \frac{f'(z)}{f(z)} = \frac{m}{z-z_0} + \frac{g'(z)}{g(z)}$$

and clearly $F(z)$ has a simple pole at z_0 with $\operatorname{Res} F(z_0) = m$. By applying this reasoning at each zero of $f(z)$ inside C we see that the sum of the residues of $F(z)$ due to the zeros of $f(z)$ inside C equals the sum, Z, of the orders of all the zeros. Thus each zero is counted according to its multiplicity in computing the sum, Z.

Similarly, if $f(z)$ has a pole of order n at z_1 then in a neighborhood of z_1 we have

$$f(z) = \frac{h(z)}{(z-z_1)^n} \quad \text{for } h(z) \text{ analytic at } z_1 \text{ and } h(z_1) \neq 0.$$

It follows that

$$F(z) = \frac{f'(z)}{f(z)} = \frac{-n}{z-z_1} + \frac{h'(z)}{h(z)}$$

and $F(z)$ has a simple pole at z_1 with $\operatorname{Res} F(z_1) = -n$. Then the sum, P, of the residues of $F(z)$ due to the poles of $f(z)$ inside C equals the sum of the orders of all the poles of f inside C. Each pole is counted according to its multiplicity in computing this sum P. Combining these two results, we have

$$\int_C F(z)\,dz = \int_C \frac{f'(z)}{f(z)}\,dz = 2\pi i(Z-P).$$

Finally note that

$$F(z) = \frac{d}{dz}\log f(z)$$

hence

$$\int_C F(z)\,dz = \Delta_C \log f(z) = \text{change in } \log f(z) \text{ as } z \text{ traces } C \text{ once in the positive sense}$$

But $\log f(z) = \ln|f(z)| + i\arg f(z)$ and since $\ln|f(z)|$ is a real valued function that returns to its original value as z traces C once, it follows that

$$\Delta_C \log f(z) = i\Delta_C \arg f(z) = 2\pi i(Z-P)$$

SUMMARY

Suppose $f(z)$ has a Laurent series expansion of the form (5.1) converging in a deleted neighborhood of $z = z_0$.

(a) *z_0 is a removable singularity for f if $b_n = 0$ for all n or, equivalently, if $f(z)$ tends to a finite limit as z approaches z_0. At a removable singularity,* $\operatorname{Res} f(z_0) = 0$.

(b) *z_0 is a simple pole if b_1 is not zero but $b_n = 0$ for $n > 1$. At any pole $f(z)$ tends to infinity as z approaches z_0. At a simple pole,*

$$\operatorname{Res} f(z_0) = \lim_{z \to z_0} (z - z_0) f(z) = b_1$$

i.e.,

$$\text{if } f(z) = \frac{g(z)}{z - z_0} \text{ for } g(z_0) \neq 0, \text{ then } \operatorname{Res} f(z_0) = g(z_0)$$

if $f(z) = p(z)/q(z)$ *where* $q(z_0) = 0,\ p(z_0) \neq 0,\ q'(z_0) \neq 0$ *then* $\operatorname{Res} f(z_0) = p(z)/q'(z)$.

(c) *z_0 is a pole or order m if b_m is not zero but $b_n = 0$ for $n > m$. At a pole of order m, $g(z) = (z - z_0)^m f(z)$ tends to a finite limit as z approaches z_0. Then*

$$\operatorname{Res} f(z_0) = \frac{1}{(m-1)!} g^{(m-1)}(z_0)$$

(d) *z_0 is an essential singularity if b_n is different from zero for infinitely many n or, equivalently, if $f(z)$ tends to no limit at z_0. In this case*

$\text{Res}f(z_0) = b_1$, *the residue can only be found from the Laurent series.*

The residue theorem can be used to evaluate a variety of real integrals. In applying the theorem the following results are often useful:

Consider the complex contour integrals

$$\int_{C_R} F(z)\,dz \quad \text{and} \quad \int_{C_R} G(z)\,e^{ibz}dz \quad \text{for } b > 0$$

where C_R *denotes the closed semicircular arc from* $(-R, 0)$ *to* $(R, 0)$ *along the real axis, and then along the semicircle* $r = Re^{i\vartheta}$, $0 < \vartheta < \pi$. *Suppose for constants* $M > 0$ *and* $p > 0$,

$$|F(z)| \le \frac{M}{R^{p+1}} \quad \text{and} \quad |G(z)| \le \frac{M}{R^p} \quad \text{for } z = Re^{i\vartheta}.$$

Then

$$\lim_{R \to \infty} \int_{C_R} F(z)\,dz = \int_{-\infty}^{\infty} F(x)\,dx$$

and

$$\lim_{R \to \infty} \int_{C_R} G(z)\,e^{ibz}dz = \int_{-\infty}^{\infty} G(x)\,e^{ibx}dx$$

The argument principle and, to a lesser extent, Rouche's theorem can be used to determine the number of zeros of an analytic function contained within a given contour. The results are most often applied to polynomials to determine, for example, if the polynomial has any zeros located in the right half plane.

SUPPLEMENTARY PROBLEMS

Use the residue theorem to evaluate the following integrals:

1. $\displaystyle\int_0^{\infty} \frac{\cos ax}{(x^2+b^2)^2}dx = \frac{\pi}{4b^3}e^{-ab}(1+ab)$ for $a > 0, b > 0$

2. $\displaystyle\int_{-\infty}^{\infty} \frac{x\sin ax}{x^2+4}dx = \frac{\pi}{2}e^{-a}\sin a$ for $a > 0$

3. $\displaystyle\int_0^{\infty} \frac{x^2}{x^6+1}dx = \pi/6$

4. $\displaystyle\int_0^{\infty} \frac{x^2}{(x^2+a^2)^2}dx = \frac{\pi}{4a}$

5. $\int_{-\infty}^{\infty} \frac{x^2+1}{x^4+1}\, dx = 2\pi/\sqrt{2}$

6. $\int_0^{\pi} \frac{d\vartheta}{(a+\cos\vartheta)^2} = \pi a(a^2-1)^{-3/2}$ for $a > 1$.

7. $\int_0^{\infty} \frac{x^{-3/4}}{x+1}\, dx = \pi\sqrt{2}$

8. $\int_0^{\infty} \frac{x^{-1/2}}{(x^2+1)^2}\, dx = \frac{3\pi}{4\sqrt{2}}$

6

Conformal Mapping

In previous chapters we have emphasized the analytic aspects of complex function theory. In this chapter we will examine some of the geometrical properties of analytic functions of a complex variable. We will see that a complex function naturally defines a mapping from the complex plane into itself and we are going to study the mapping properties of analytic functions in particular.

We begin this chapter by recalling the notion of a Jacobian determinant for a mapping of the plane into itself and we see that for a mapping defined by an analytic function, $w = F(z)$, the Jacobian is related to the derivative, $F'(z)$. Particular mapping properties of several elementary analytic functions are then examined.

We also show that a harmonic function $\varphi = \varphi(x, y)$ remains harmonic under a transformation $(x, y) \to (u, v)$, defined by an analytic function $w = F(z)$. This fact can be used for solving boundary value problems for Laplace's equation. The various analytic mappings studied in this Chapter will also be applied in Chapter 8 in connection with our development of potential theory.

MAPPINGS IN THE PLANE

Consider a pair of real valued functions $u = u(x, y)$, $v = v(x, y)$ defined and smooth in an open set Ω in the xy-plane. These functions can be viewed as defining a *mapping* T, of the plane into itself; i.e., T maps the point (x, y) into the point $(u(x, y), v(x, y))$. As the point $P = (x, y)$ traces out a curve C in the xy-plane, the *image point* $Q = (u, v)$ traces out a corresponding *image curve* C' in the uv-plane. Similarly, if P sweeps out a region D in the xy-plane, then $Q = (u, v)$ sweeps out a region D' which is the *image of D under the mapping T.*

THE JACOBIAN OF A MAPPING

Let the mapping T be defined by smooth functions u and v

$$T: \quad \begin{matrix} u = u(x,y) \\ v = v(x,y) \end{matrix} \qquad \text{for } (x,y) \text{ in } \Omega$$

and define the *Jacobian of the mapping* T at the point (x_0, y_0) to be the following determinant

$$J(x_0, y_0) = \det \begin{bmatrix} \partial_x u(x_0, y_0) & \partial_y u(x_0, y_0) \\ \partial_x v(x_0, y_0) & \partial_y v(x_0, y_0) \end{bmatrix} = \partial_x u \partial_y v - \partial_y u \partial_x v \qquad (6.1)$$

We will also use the following alternative notation for the Jacobian of T

$$J = \frac{\partial(u, v)}{\partial(x, y)}.$$

The importance of the Jacobian lies in the fact that the mapping T is one to one in a neighborhood of any point where its Jacobian, J, is not equal to zero.

Theorem 6.1 Suppose real valued functions $u = f(x, y)$ and $v = g(x, y)$ are defined and continuously differentiable on an open set Ω in the plane. Suppose that at (x_0, y_0) in Ω we have $u_0 = f(x_0, y_0)$, $v_0 = g(x_0, y_0)$ and $J(x_0, y_0) \neq 0$. Then there exist positive numbers α and β which determine two boxes

$$B = \{ (x,y) : |x - x_0| < \alpha, |y - y_0| < \alpha \}$$

and

$$B' = \{ (u, v) : |u - u_0| < \beta, |v - v_0| < \beta \}$$

such that there is a one to one correspondence between points (u, v) in B' and points (x, y) in B; i.e., for each (u, v) in B' there exists a unique (x,y) in B with $(u, v) = (f(u, v), g(x, y))$ and, since the correspondence is one to one, the functions f and g induce inverse functions f^{-1} and g^{-1} such that for each (x, y) in B there is a unique (u, v) in B' such that $(x, y) = (f^{-1}(u, v), g^{-1}(u, v))$.

In the solved problems we show that the Jacobian J can be interpreted as the ratio of the area of B' to the area of B (i.e., J is a local scale factor for the mapping).

ANALYTIC FUNCTIONS AS MAPPINGS

Consider the complex function $w = F(z) = u(x, y) + iv(x, y)$, defined and analytic on a domain Ω in the z-plane. Then F defines a mapping from the z-plane to the w-plane and since F is analytic we have

$$
\begin{aligned}
J &= \partial_x u \partial_y v - \partial_y u \partial_x v \\
&= (\partial_x u)^2 + (\partial_y u)^2 = (\partial_x v)^2 + (\partial_y v)^2 = |F'(z)|^2
\end{aligned}
\tag{6.2}
$$

Thus by Theorem 6.1 the mapping F is one to one in a neighborhood of any point z_0 where $F'(z_0)$ is not zero; i.e., there is a one to one correspondence between points $z = (x, y)$ in a neighborhood of $z_0 = (x_0, y_0)$, and points $w = (u, v)$ in a neighborhood of $w_0 = F(z_0)$ so that

$$w = F(z) = u(x, y) + iv(x, y)$$

and

$$z = F^{-1}(w) = x(u, v) + iy(u, v)$$

with

$$
\begin{aligned}
|dF/dz|^2 &= (\partial_x u)^2 + (\partial_y u)^2 = (\partial_x v)^2 + (\partial_y v)^2 = J \\
|dF^{-1}/dz|^2 &= (\partial_x u)^2 + (\partial_y u)^2 = (\partial_x v)^2 + (\partial_y v)^2 = 1/J
\end{aligned}
\tag{6.3}
$$

It now follows that a harmonic function remains harmonic under an analytic transformation (change of variables).

Theorem 6.2 Let $w = F(z)$ be defined and analytic on a domain Ω in the z-plane and suppose $\varphi = \varphi(x, y)$ is a real valued function defined and harmonic on Ω. Then $\Phi(u, v) = \varphi(x(u, v), y(u, v))$ is harmonic on Ω', the image of Ω under the mapping F.

CONFORMAL MAPPINGS

Let C_1, C_2 be two smooth curves in the z-plane intersecting at the point z_0. Let L_1 and L_2 denote lines tangent to C_1, C_2 at z_0 and let ϑ denote the angle between the two tangents, measured from L_1 to L_2. Suppose $F(z)$ is a function that is analytic in a domain Ω containing C_1 and C_2 and let C_1', C_2' denote the images of C_1, C_2 under the mapping defined by F. Then C_1', C_2' intersect at the point $w_0 = F(z_0)$ and have tangents L_1' and L_2' at w_0. If $F'(z_0)$ is not zero then we can show that L_1' and L_2' are obtained by rotating L_1 and L_2 through the same angle $\alpha = \arg F'(z_0)$. Then the angle between the tangents measured from L_1' to L_2' is again equal to ϑ. A mapping which preserves the sense and magnitude of angles between curves is said to be a *conformal mapping*.

Theorem 6.3 Suppose $F(z)$ is analytic on the domain D. Then the mapping $F(z)$ is conformal at each point z in D where $F'(z)$ is not zero.

Note that for a conformal mapping F, $|F'(z)|$ can be interpreted as the local magnification factor for the mapping at z and $\arg F'(z)$ represents the angle by which the inclination of the tangent to a curve through z is increased by the mapping; i.e., if C is a curve through z and L is tangent to C at z then the image curve C' has tangent L' at $w = F(z)$ where the inclination of L' exceeds that of L by the amount $\arg F'(z)$.

THE POINT AT INFINITY

It will be convenient to have the notion of the complex *point at infinity.* To define this concept consider a sphere of radius one with its center at the origin in the complex plane. Then the complex plane is the equatorial plane of this sphere and we denote by P the point at the north pole of the sphere. For each z in the complex plane the line joining P to z cuts the sphere at a unique point $Q(z)$ and it is easy to see that:

(i) if $|z| > 1$ then $Q(z)$ lies in the northern hemisphere
(ii) if $|z| < 1$ then $Q(z)$ lies in the southern hemisphere
(iii) if $z = 0$ then $Q(z)$ is the south pole of the sphere

There is a one to one correspondence between the points of the complex plane and the points of the sphere excluding only P. Then we associate the *point at infinity* in the complex plane with the point P on the sphere. We refer to this sphere as the *Riemann sphere* and to the set of points corresponding to all the points of the sphere as the *extended complex plane.* The point of the plane associated with the point P on the sphere is denoted by ∞. This symbol is not to be used in arithmetic operations but will be convenient for certain purposes in what follows.

ELEMENTARY FUNCTIONS AS MAPPINGS

LINEAR MAPPINGS

Consider the complex linear function $F(z) = az + b$ for given complex constants a and b. This mapping is one to one at all points if a and b are not both zero.

EXAMPLE 6.1

(a) The mapping $w = 2e^{i\pi/4}z$ maps the unit square $OABC$ in the z-plane into the square $O'A'B'C'$ in the w-plane as shown in Figure 6.1.

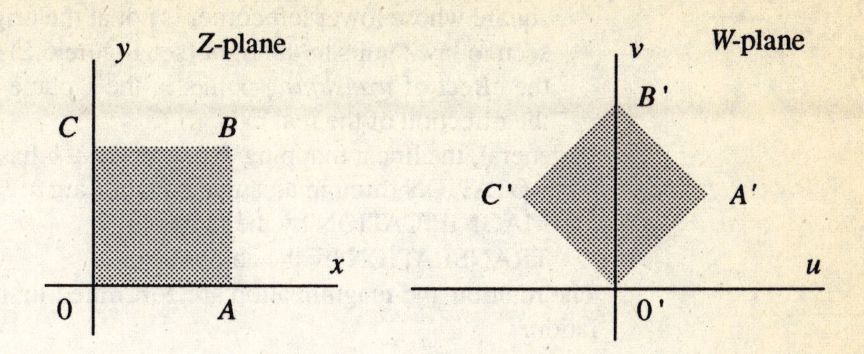

Figure 6.1. Mapping $F(z) = Az$

To see this note that

point O, $z = 0$, is mapped to O', $F(0) = 0$

point A, $z = 1 = e^{i0}$, is mapped to A', $F(1) = 2e^{i\pi/4}e^{i0} = \sqrt{2} + i\sqrt{2}$

point B, $z = 1 + i = \sqrt{2}e^{i\pi/4}$, is mapped to B', $F(\sqrt{2}e^{i\pi/4}) = 2\sqrt{2}e^{i\pi/2} = 2\sqrt{2}i$

point C, $z = i = e^{i\pi/2}$, is mapped to C', $F(i) = 2e^{i3\pi/4} = -\sqrt{2} + i\sqrt{2}$

In addition, it is easy to see that points on the segment OA map onto points lying on the segment $O'A'$ just as points on the other sides of the square $OABC$ map onto points lying on the corresponding sides of $O'A'B'C'$. Finally, points in the interior of $OABC$ go into the interor of $O'A'B'C'$. Thus the image of the square is seen to be a square that has been *rotated* through $\pi/4$ radians and has been *magnified* by a factor of 2.

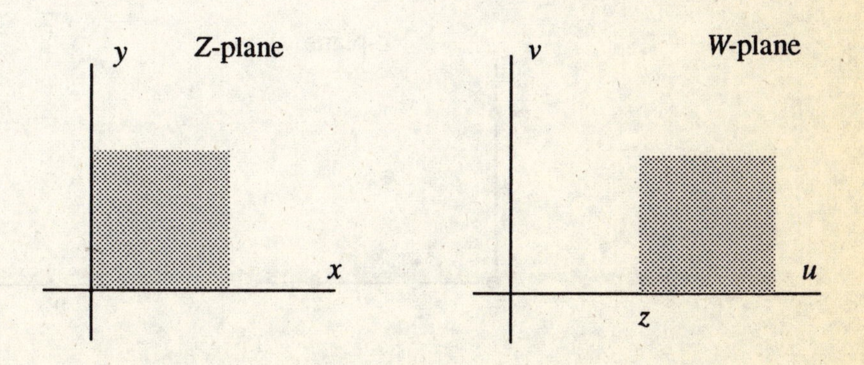

Figure 6.2. Mapping $F(z) = z + 2$, Translation

(b) The linear mapping $w = z+2$ maps the unit square onto a similar square whose lower left corner is not at the origin in the w-plane but is seen to lay 2 units to the right (see Figure 6.2). Thus this mapping has the effect of *translating* points in the z-plane a distance of 2 units in the direction of the real axis.

In general, the linear mapping $F(z) = az+b$ has the following effects:

ROTATION through an angle equal to $\arg a$

MAGNIFICATION by the factor $|a|$

TRANSLATION by the amount b

The rotation and magnification are performed first, followed by the translation.

THE QUADRATIC MAPPING

Consider the function $F(z) = z^2 = x^2 - y^2 + i2xy$. This mapping is one to one in a neighborhood of every point except $z = 0$. Note that F can also be expressed in polar notation as

$$F(re^{i\vartheta}) = r^2 e^{i2\vartheta}.$$

EXAMPLE 6.2

(a) The mapping F carries the first quadrant in the z-plane onto the upper half of the w-plane as shown in Figure 6.3. Clearly F carries $z = 0$ onto $w = 0$ and any point $z = re^{i0} = r$ on the positive real axis is mapped onto $w = r^2 e^{i0} = r^2$ on the positive real axis in the w-plane. Similarly, any point $z = re^{i\pi/2}$ on the positive imaginary axis in the z-plane maps onto $w = r^2 e^{i\pi}$ which lies on the negative real axis in the w-plane. Finally note that any point $z = re^{i\vartheta}$, $0 < \vartheta < \pi/2$, in the interior of the quarter plane is mapped by F to a point $w = r^2 e^{i2\vartheta}$ in the interior of the upper half of the w-plane.

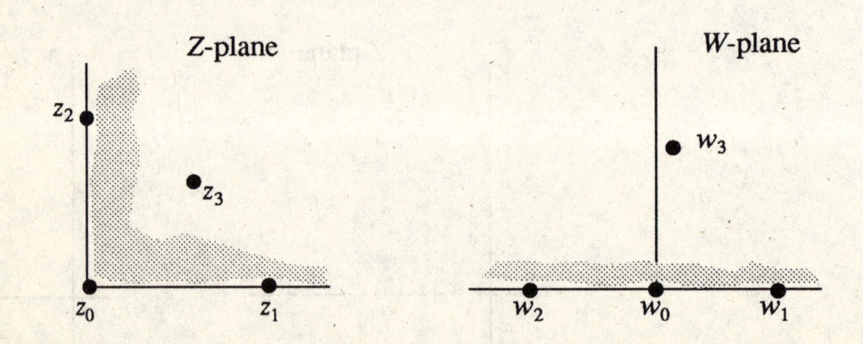

Figure 6.3. Mapping $F(z) = z^2$

(b) In the same way we can show that the upper half of the z-plane is mapped by F onto the whole w-plane. We can also show that the whole z-plane is mapped onto the (doubly covered) w-plane. Each point in the w-plane is the image of two points in the z-plane. This does not violate Theorem 6.1 which implies only that the mapping F is one to one *in a neighborhood* of each point where $F'(z)$ is not zero. We say then that F is *locally one to one* at each point in the z-plane except $z = 0$ which is a branch point for the inverse mapping $F^{-1}(w) = w^{1/2}$.

(c) Consider the two strips in the z-plane $S_1 = \{ (x,y) : 1 \le xy \le 2 \}$ and $S_2 = \{ (x,y) : -2 \le x^2 - y^2 \le -1 \}$. These regions are shown in Figure 6.4 where they can be seen to be bounded by members of the orthogonal families of hyperbolas $xy =$ Constant and $x^2 - y^2 =$ Constant, respectively.

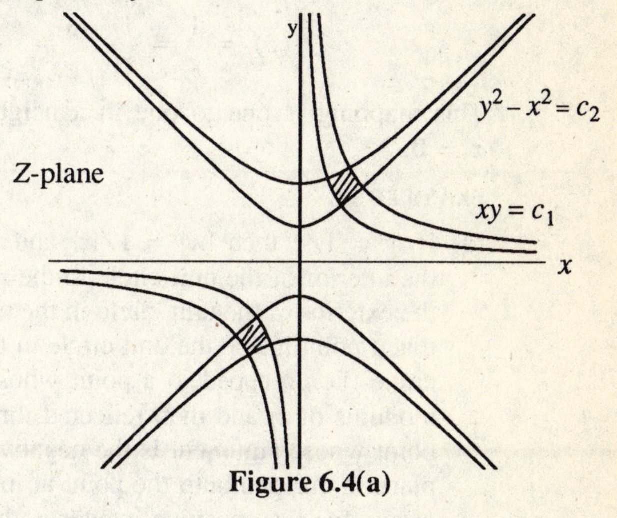

Figure 6.4(a)

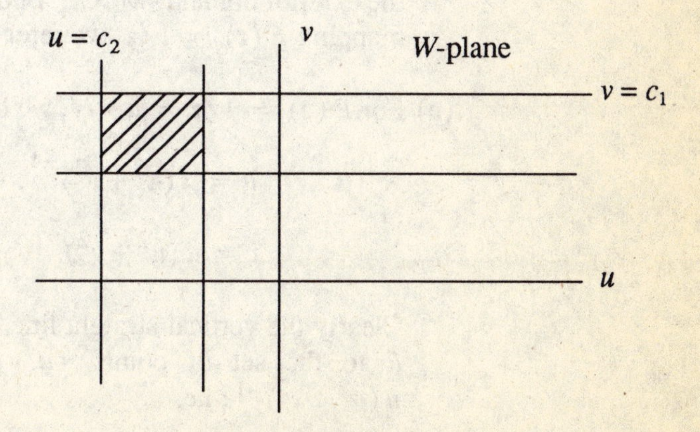

Figure 6.4(b)

These regions are mapped by F onto the strips $S'_1 = \{(u,v): 1 \le v \le 2\}$ and $S'_2 = \{(u,v): -2 \le u \le -1\}$ in the w-plane. Note that the hyperbolas of the form $xy = C_1$ are orthogonal in the z-plane to the hyperbolas $x^2 - y^2 = C_2$. These curves are mapped onto lines $v = C_1'$ and $u = C_2'$ which are likewise orthogonal to each other in the w-plane. The angle between intersecting curves in the z-plane is preserved by the mapping F as predicted by Theorem 6.3. Note also that the branches of the hyperbolas lying in the lower half of the z-plane are mapped on top of the w-plane images of the upper branches of the hyperbolas. This illustrates the two to one nature of the mapping $F(z) = z^2$.

THE INVERSION MAPPING

Consider the complex function

$$F(z) = \frac{1}{z} = \frac{1}{r}e^{-i\vartheta} = \frac{x}{x^2+y^2} - i\frac{y}{x^2+y^2}$$

This mapping is one to one in a neighborhood of every point except $z = 0$.

EXAMPLE 6.3

(a) If $w = 1/z$ then $|w| = 1/|z|$ and $\arg w = -\arg z$. Thus points in the interior of the unit circle in the z-plane, $|z| < 1$, are mapped onto the exterior of the unit circle in the w-plane, $|w| > 1$. Note, however, that a point inside the unit circle in the z-plane is first inverted in the circle (i.e., mapped to a point whose modulus is the inverse of the modulus of z) and then reflected through the real axis (mapped to a point whose argument is the negative of $\arg z$). The origin in the z-plane is mapped onto the point at infinity in the w-plane. More precisely, for each positive integer n, the disc $|z| < 1/n$ is mapped onto the exterior domain $|w| > n$. Because of this inversion property of the mapping $F(z) = 1/z$, we refer to F as the *inversion mapping*.

(b) For $F(z) = 1/z = u + iv$, we have $u^2 + v^2 = (x^2+y^2)^{-1}$ and

$$u = x(x^2+y^2)^{-1}, \; v = -y(x^2+y^2)^{-1}$$

or

$$x = u(u^2+v^2)^{-1}, \; y = -v(u^2+v^2)^{-1}.$$

Clearly the vertical straight line, $x = C$, in the z-plane is mapped by F to the set of points (u,v) in the w-plane for which $C = u(u^2+v^2)^{-1}$; i.e.

$$u^2 + v^2 - u/C = 0 \quad \text{or} \quad (u - 1/2C)^2 + v^2 = (1/2C)^2.$$

This is the equation of a circle in the w-plane having center at $(1/2C, 0)$ on the u-axis, and radius equal to $1/2C$. This circle is tangent to the v-axis at the origin. Note that the points lying to the *right* of the vertical line in the z-plane (i.e., $x > C$) get mapped onto the *interior* of the circle while the points to the *left* of the line are mapped onto the domain *outside* the circle.

Similarly, the horizontal straight line $y = D$ in the z-plane is mapped by F onto the set of points (u, v) in the w-plane for which $D = -v(u^2 + v^2)^{-1}$. These points satisfy the equation

$$u^2 + v^2 + v/D = 0 \quad \text{or} \quad (v + 1/2D)^2 + u^2 = (1/2D)^2.$$

Thus the image of the line $y = D$ is the circle of radius $1/2D$ in the w-plane whose center is at $(0, -1/2D)$ on the v-axis. This circle is tangent to the u-axis at the origin. The half-plane lying *above* the horizontal line $y = D$ is transformed by F to the *interior* of the circle.

It is evident from Figure 6.5 that the families of lines $x = C$ and $y = D$ are orthogonal families in the z-plane and the corresponding circles in the w-plane are also orthogonal families, illustrating that $F(z) = 1/z$ is a conformal mapping.

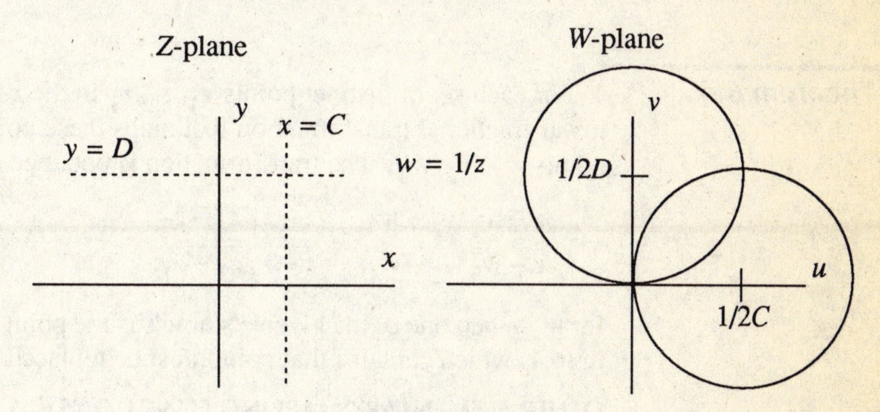

Figure 6.5

(c) By reasoning similar to that just applied in part (b), we can show that circles in the z-plane with center on the x-axis and tangent at the origin to the y-axis are mapped by F to horizontal straight lines in the w-plane. Likewise, circles whose center is on the y-axis and which pass through the origin are mapped by F onto vertical straight lines in the w-plane. We can summarize these observations by saying that $F(z) = 1/z$ carries lines parallel to the axes into circles through the origin and circles through the origin into lines parallel to the axes. If we think of lines as circles through the point at infinity then we can say the inversion mapping carries circles through the origin to circles through the point at infinity and vice versa.

LINEAR FRACTIONAL TRANSFORMATIONS

Consider the complex function

$$F(z) = \frac{az+b}{cz+d} = \frac{a}{c} + \frac{bc-ad}{c}\frac{1}{cz+d}$$

where a, b, c, d are complex constants with $ad - bc \neq 0$. Then we refer to the mapping $w = F(z)$ as a *linear fractional transformation.* If $w = F(z)$ we have

$$z = F^{-1}(w) = \frac{-dw+b}{cw-a}$$

and it follows that there is a one to one correspondence between points in the extended z-plane and points in the extended w-plane. In particular, for c different from zero, $z = -d/c$ is mapped to $w = \infty$ and $z = \infty$ is mapped to $w = a/c$. If $c = 0$ then $z = \infty$ corresponds to $w = \infty$.

If we think of straight lines as circles of infinite radius then we can show that all linear fractional transformations map circles into circles. The following result is often useful.

Theorem 6.4 For each set of distinct points z_1, z_2, z_3 in the z-plane there is a unique linear fractional transformation that maps these points onto given distinct points w_1, w_2, w_3. The transformation is obtained by solving

$$\frac{w-w_1}{w-w_3}\frac{w_2-w_3}{w_2-w_1} = \frac{z-z_1}{z-z_3}\frac{z_2-z_3}{z_2-z_1} \tag{6.4}$$

for w. When one of the given z's or w's is the point at infinity, the quotient in (6.4) which contains that point must be replaced by 1.

OTHER ELEMENTARY MAPPING FUNCTIONS

Other elementary functions which produce interesting mappings to be considered in the solved problems include:

$F(z) = z^p$ for p not necessarily an integer

$F(z) = e^{bz}$ for complex constant b

$F(z) = \sin z, \cos z$

$F(z) = z + 1/z$

SUCCESSIVE TRANSFORMATIONS

We may also consider transformations of the form $Z = F(z)$,

$w = G(Z)$ where the transformations F and G are applied in succession.

EXAMPLE 6.4

In Problem 6.8 it is shown that the mapping $Z = F(z) = e^{\pi z/a}$ for $a > 0$ maps the strip $S = \{0 < \operatorname{Im} z < a\}$ onto the upper half of the Z-plane. In Problem 6.3 we show that the mapping

$$w = G(Z) = \frac{Z-i}{Z+i}$$

takes the uper half of the Z-plane onto $|w| < 1$. Then the compound mapping

$$w = G(F(z)) = \frac{e^{\pi z/a} - i}{e^{\pi z/a} + i}$$

maps the strip S onto the interior of the unit disc in the w-plane.

THE RIEMANN MAPPING THEOREM

We see in this chapter, several examples of mappings of various domains in the z-plane onto the interior of the unit disc in the w-plane. The famous *Riemann mapping theorem* states that for any simply connected domain Ω which is not the whole z-plane, there exists an analytic function $F(z)$ which defines a one to one mapping of Ω onto the disc $|w| < 1$. The theorem provides no information on how the mapping F is to be found. However, if Ω is bounded and simply connected with a simple closed polygonal boundary, there is a general method for finding the mapping F. The method, known as the *Schwarz-Christoffel formula,* involves integrals that frequently lead to nonelementary functions. Further discussion of these two topics is beyond the scope of this text.

SOLVED PROBLEMS

Mappings of the Plane

PROBLEM 6.1

Suppose T is a continuously differentiable mapping from the plane into itself and that the Jacobian J of the mapping is nonzero at (x_0, y_0). For $\Omega = \{|x - x_0| < \varepsilon, |y - y_0| < \varepsilon\}$ show that

$$\frac{|T(\Omega)|}{|\Omega|} \to |J(x_0, y_0)| \quad \text{as} \quad \varepsilon \to 0 \tag{1}$$

where $|\Omega|$ denotes the area of the domain Ω and $|T(\Omega)|$ denotes the area of its image under T.

SOLUTION 6.1

By definition

$$|T(\Omega)| = \iint_{T(\Omega)} 1 du dv$$

where 1 denotes the function that is identically equal to one. Then by the rule for changing variables in a double integral,

$$|T(\Omega)| = \iint_{T(\Omega)} 1 du dv = \iint_{T(\Omega)} 1 \left| \frac{\partial(u, v)}{\partial(x, y)} \right| dx dy = \iint_{\Omega} |J| dx dy \quad (2)$$

Then the mean value theorem for integrals applied to (2), implies

$$|T(\Omega)| = \iint_{\Omega} |J| dx dy = |J(x, y)| \iint_{\Omega} dx dy = |J(x, y)| |\Omega| \quad (3)$$

for some point (x, y) in $\Omega = \{|x - x_0| < \varepsilon, |y - y_0| < \varepsilon\}$. Since J is assumed to be continuous, it follows from (3) that

$$\frac{|T(\Omega)|}{|\Omega|} = |J(x, y)| \to |J(x_0, y_0)| \quad \text{as } \varepsilon \text{ tends to zero.}$$

PROBLEM 6.2

Let $w = F(z)$ be defined, analytic and one to one on a domain Ω in the z-plane and suppose $\varphi = \varphi(x, y)$ is a real valued function defined and harmonic on Ω. Then show that:

(a) $\Phi(u, v) = \varphi(x(u, v), y(u, v))$ is harmonic on Ω', the image of Ω under F

(b) if $\varphi = A$ (constant) along curve C in Ω, then $\Phi = A$ along $C' = F(C)$ in Ω'

(c) if the normal derivative $\partial_n \varphi$ vanishes along curve C in Ω, then $\partial_n \Phi = 0$ along the image curve C' in Ω'

SOLUTION 6.2

To show (a) we use the chain rule to differentiate $\Phi(u, v) = \varphi(x(u, v), y(u, v))$

$$\partial_u \Phi = \partial_x \varphi \partial_u x + \partial_y \varphi \partial_u y, \quad \partial_v \Phi = \partial_x \varphi \partial_v x + \partial_y \varphi \partial_v y$$

$$\partial_{uu} \Phi = \partial_{xx} \varphi (\partial_u x)^2 + 2\partial_{xy} \varphi \partial_u x \partial_u y + \partial_{yy} \varphi (\partial_u y)^2 + \partial_x \varphi \partial_{uu} x + \partial_y \varphi \partial_{uu} y$$

$$\partial_{vv} \Phi = \partial_{xx} \varphi (\partial_v x)^2 + 2\partial_{xy} \varphi \partial_v x \partial_v y + \partial_{yy} \varphi (\partial_v y)^2 + \partial_x \varphi \partial_{vv} x + \partial_y \varphi \partial_{vv} y$$

Then

$$\partial_{uu}\Phi+\partial_{vv}\Phi = \partial_{uu}\varphi((\partial_u x)^2+(\partial_v x)^2)+2(\partial_{xx}\varphi(\partial_u x\partial_u y+\partial_v x\partial_v y))$$

$$+\partial_{yy}\varphi((\partial_u y)^2+(\partial_v y)^2)+\partial_x\varphi(\partial_{uu}x+\partial_{vv}x)+\partial_y\varphi(\partial_{uu}y+\partial_{vv}y).$$

But (6.3) implies $(\partial_u x)^2+(\partial_v x)^2 = (\partial_u y)^2+(\partial_v y)^2 = 1/J$, and it follows from the Cauchy Riemann equations applied to $x = x(u,v)$ and $y = y(u,v)$ that $\partial_u x = \partial_v y$ and $\partial_v x = -\partial_u y$. Finally, since $x = x(u,v)$ and $y = y(u,v)$ are, respectively, the real and imaginary parts of the analytic function F^{-1} we have $\partial_{uu}x+\partial_{vv}x = 0$ and $\partial_{uu}y+\partial_{vv}y = 0$. Then $\partial_{uu}\Phi+\partial_{vv}\Phi = (\partial_{xx}\varphi+\partial_{yy}\varphi)/J$ and Φ is harmonic at all points where φ is harmonic and J is different from zero.

Now suppose $\varphi = A$ (constant) along curve C in Ω and let C' denote the image of C under the mapping F. Then as w traces C', $z = x(u,v)+iy(u,v)$ traces C and it follows that $\Phi(u,v) = \varphi(x(u,v),y(u,v)) = A$ along $C' = F(C)$ in Ω'. This proves (b).

Finally, suppose $\partial_n\varphi$, the normal derivative of φ, vanishes on C. Then it follows (see Problem 2.10) that C is a level curve for φ; i.e., φ = constant along C. But by the result (b) then Φ = constant along C'. But if Φ = constant along C', then $\partial_n\Phi = 0$ along C' and this proves (c). Note that (c) also follows from the fact that $\partial_n\Phi(u,v) = \partial_n\varphi(x,y)/J$.

Analytic Functions as Mappings

PROBLEM 6.3

Show that for z_0 given with $\text{Im}\, z_0 > 0$, the linear fractional transformation

$$w = \frac{z-z_0}{z-z_0^*} \tag{1}$$

maps the upper half plane onto the interior of the unit circle in the w-plane with z_0 going to $w_0 = 0$, the origin in the w-plane.

SOLUTION 6.3

Note that the set of points z such that $|z-z_0| = |z-z_0^*|$ is the set of points in the z-plane that are equidistant from $z_0 = x_0+iy_0$ and its complex conjugate $z_0^* = x_0-iy_0$. But this is just the points of the real axis and it follows then from (1) that the x-axis in the z-plane is mapped onto $|w| = 1$, the unit circle in the w-plane. Now the set of points z such that $|z-z_0| < |z-z_0^*|$ is the set of points in the z-plane that are closer to z_0 than to its complex conjugate z_0^*; i.e., if $\text{Im}\, z_0 > 0$, this is the upper half of the z-plane. Then (1) implies this set is mapped onto $|w| < 1$, the interior of the unit circle in the w-plane. Finally, (1) implies that when $z = z_0$ then $w = 0$ so z_0 is mapped to the origin.

In the same way we can show that for positive real number a,

$$w = \frac{z-a}{z+a} \tag{2}$$

maps the right half of the z-plane into the interior of the unit circle in the w-plane with $z = a$ going to the origin.

PROBLEM 6.4

Show that for each set of distinct points z_1, z_2, z_3 in the z-plane there is a unique linear fractional transformation that maps these points onto given distinct points w_1, w_2, w_3 in the w-plane.

SOLUTION 6.4

Suppose that

$$F(z) = \frac{az+b}{cz+d} \tag{1}$$

maps four distinct points z_1, z_2, z_3, z_4 in the z-plane onto image points w_1, w_2, w_3, w_4 respectively. Using (1) we find that if all of the points are finite then for $1 \leq j, k \leq 4$

$$w_j - w_k = F(z_j) - F(z_k) = \frac{ad-bc}{(cz_j+d)(cz_k+d)}(z_j - z_k) \tag{2}$$

for j different from k. There are four distinct equations in (2) corresponding to various choices of j different from k. By forming quotients of these equations it follows that

$$\frac{w_4-w_1}{w_4-w_3}\frac{w_2-w_3}{w_2-w_1} = \frac{z_4-z_1}{z_4-z_3}\frac{z_2-z_3}{z_2-z_1} \tag{3}$$

The product of quotients on the right in (3) is called the *cross ratio* of the points z_1 to z_4. Then (3) implies that the cross ratio of the points z_1 to z_4 is the same as the cross ratio of their images. Since (3) does not depend on the numbers a, b, c, d this must be true for all linear fractional transformations. Since (3) holds for all finite points z_1 to z_4 and their finite images, we can replace z_4 and w_4 in (3) by variables z and w to obtain (6.4). Then solving (6.4) for w in terms of z yields the unique linear fractional transformation that carries z_1, z_2, z_3 into w_1, w_2, w_3.

When one of the given z's or w's in (3) is the point at infinity, we divide the numerator and denominator of the quotient in which it appears by this particular z or w. Letting this z or w tend to infinity causes that quotient to tend to the value 1. Thus the quotient in (6.4) which contains an infinite z or w must be replaced by 1.

PROBLEM 6.5

Find the linear fractional transformation T that maps the points $z_1 = 0$, $z_2 = 1+i$, $z_3 = 2$ from the circle $|z-1| = 1$ onto the points $w_1 = -i$, $w_2 = 1$ and $w_3 = i$ on the unit circle in the w-plane. Show that T maps the right half of the z-plane with the disc $D = \{|z-1| < 1\}$ removed, onto the interior of the unit disc in the w-plane (see Figure 6.6).

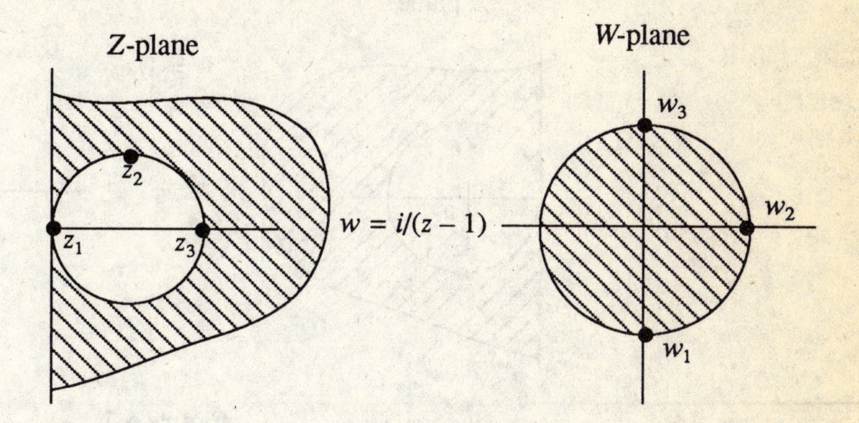

Figure 6.6

SOLUTION 6.5

We use (6.4) from Theorem 6.4 to write

$$\frac{w+i}{w-i}\frac{1-i}{1+i} = \frac{z-0}{z-2}\frac{1+i-2}{1+i-0}; \tag{1}$$

i.e.,

$$\frac{w+i}{w-i} = \frac{-z}{z-2}.$$

Solving for w in terms of z leads to $w = i/(z-1)$. It is easy to check that this mapping carries 0, $1+i$, 2 from the z-plane to $-i$, 1 and i in the w-plane. The fact that the points 0, $1+i$, 2 go around the circle $|z-1| = 1$ in the clockwise sense while $-i$, 1, i go around $|w| = 1$ in the counterclockwise sense suggest that it is the *exterior* of $|z-1| = 1$ that is mapped into the *interior* of $|w| = 1$. To check this, note that points of the form $z_p = 1+p(1+i)$ lie outside $|z-1| = 1$ for $p > 1$ and are mapped onto points of the form $w_p = (1-i)/2p$ which lie inside $|w| = 1$. Similarly $z_q = 1+q(1-i)$ lies outside $|z-1| = 1$ for $q > 1$ and is mapped to $w_q = (-1+i)/2q$ inside $|w| = 1$. Note that z_p in the upper half of the z-plane goes to w_p in the lower half of the w-plane, while z_q goes from the lower half z-plane to w_q in the upper half of the w-plane.

PROBLEM 6.6

Show that for positive real number a, the inversion mapping $F(z) = 1/z$ maps the right half of the z-plane exterior to the circle $|z-a| = a$ onto the strip $0 < \mathrm{Re}\, w < 1/2a$ in the w-plane (see Figure 6.7).

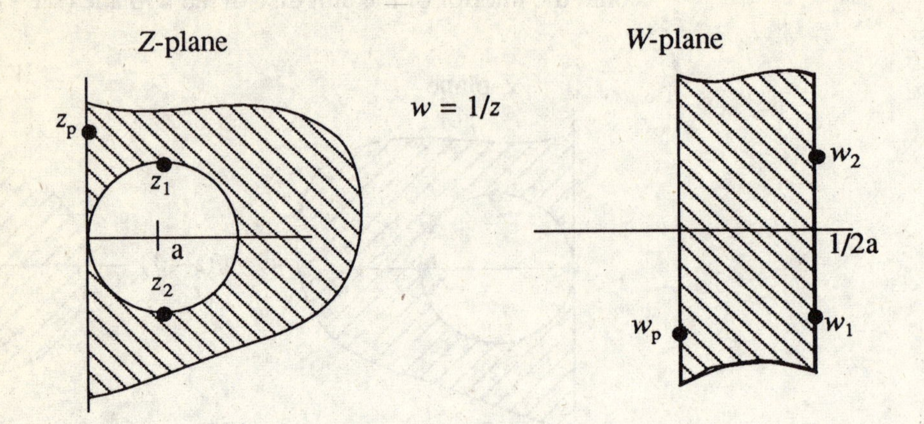

Figure 6.7

SOLUTION 6.6

Note that in general the circle $(x-a)^2 + (y-b)^2 = c^2$ in the z-plane can also be represented by the equation

$$zz^* + Az + A^*z^* + B = 0 \text{ for } A = -a+ib, B = a^2+b^2-c^2 \quad (1)$$

The equation (1) results from substituting $zz^* = x^2+y^2$, $x = (z+z^*)/2$, $y = (z-z^*)/2$ into the original equation for the circle and simplifying. Then the inversion mapping $w = 1/z$ maps the circle C to the curve

$$C'\text{: } \frac{1}{ww^*} + \frac{A}{w} + \frac{A^*}{w^*} + B = 0$$

i.e.,
$$1 + Aw^* + A^*w + Bww^* = 0 \quad (2)$$

Note that (2) has the same form as (1) consistent with our previous observation that the inversion map carries circles to circles.

The circle $|z-a| = a$ has the equation, $(x-a)^2+y^2 = a^2$; i.e., $b = 0, A = -a, B = 0$. Then (2) reduces to $1-2au = 0$ which is the equation of a vertical straight line through the point $(1/2a, 0)$ in the w-plane. This straight line is the image of the circle $|z-a| = a$ under the inversion mapping. Note that $z_1 = a(1+i)$ on the upper semicircle is sent to the point $w_1 = (1-i)/2a$ on the lower half of the vertical line and $z_2 = a(1-i)$ on the lower semicircle is mapped to the point

$w_2 = (1+i)/2a$ on the upper half of the vertical line. Note also that points $z_p = pi$ on the imaginary axis in the z-plane are mapped to $w_p = -i/p$ on the imaginary axis in the w-plane but again the points are mapped to the opposite side of the real axis. In particular, the origin in the w-plane is the image of the point at infinity in the z-plane. It is not hard to check that points on various lines and curves in the exterior of the circle $|z-a| = a$ in the z-plane are mapped into the interior of the strip $0 < \operatorname{Re} w < 1/2a$ in the w-plane.

PROBLEM 6.7

Show that the mapping $w = z^{2/3}$ maps the region $\Omega = \{z = re^{i\vartheta}: r > 0, 0 < \vartheta < 3\pi/2\}$ onto the upper half of the w-plane (see Figure 6.8).

SOLUTION 6.7

Here we take the branch cut of the multiple valued mapping function $F(z) = z^{2/3}$ to lie on the positive real axis in the z-plane. Then the z-plane rays indicated in Figure 6.8 are as follows:

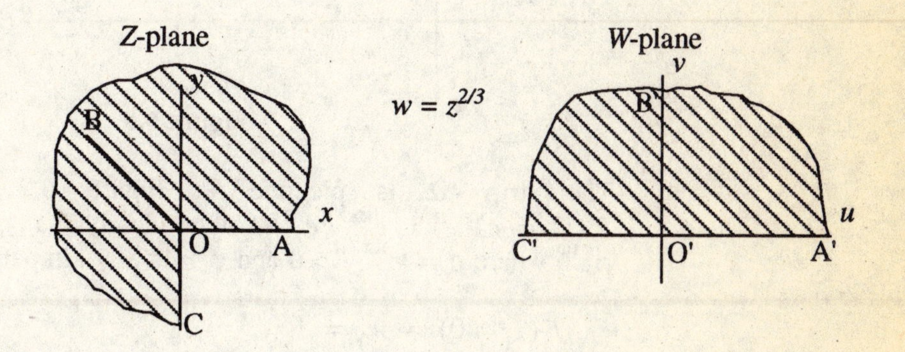

Figure 6.8

ray $OA = \{z = re^{i\vartheta}: r > 0, \vartheta = 0\}$, ray $OB = \{z = re^{i\vartheta}: r > 0, \vartheta = 3\pi/4\}$ and ray $OC = \{z = re^{i\vartheta}: r > 0, \vartheta = 3\pi/2\}$. Then F maps these rays onto corresponding rays in the w-plane:

$$\text{ray } O'A' = \{w = \rho e^{i\varphi}: \rho = r^{2/3} > 0, \varphi = 2\vartheta/3 = 0\}$$

$$\text{ray } O'B' = \{w = \rho e^{i\varphi}: \rho > 0, \varphi = \pi/2\}$$

$$\text{ray } O'C' = \{w = \rho e^{i\varphi}: \rho > 0, \varphi = \pi\}$$

In the same way it follows that F maps the entire z-plane with the branch cut on the positive real axis onto the sectorial region $\{w = \rho e^{i\varphi}: \rho > 0, 0 < \varphi < 4\pi/3\}$. In general the mapping $F(z) = z^p$,

$p > 0$, maps the infinite wedge $\{z = re^{i\vartheta}:\ r > 0, 0 < \vartheta < \alpha\}$ whose vertex angle is α onto the wedge $\{w = \rho e^{i\varphi}:\ \rho = r^p, 0 < \varphi < p\alpha\}$. Thus choosing $p = \pi/\alpha$ causes F to map the wedge onto the upper half of the w-plane.

PROBLEM 6.8

Show that for positive real number a, the mapping $w = e^{\pi z/a}$ sends the infinite strip $\Omega = \{z:\ 0 < \text{Im}\, z < a\}$ onto the upper half of the w-plane.

SOLUTION 6.8

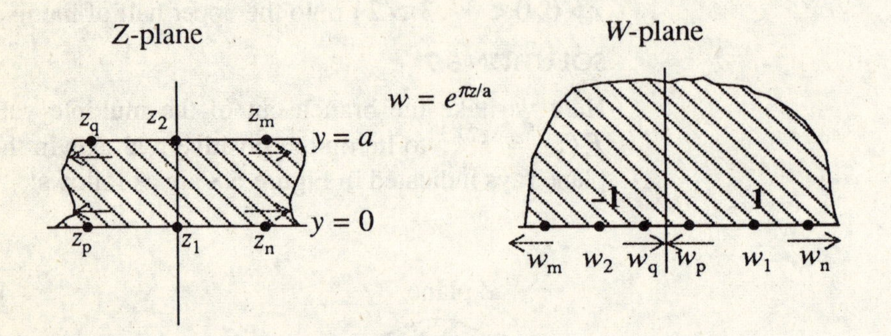

Figure 6.9

The strip Ω is pictured in Figure 6.9 and if we write $F(z) = e^{\pi z/a} = e^{\pi x/a} e^{i\pi y/a}$ then clearly F maps $z = x + iy$ to $w = \rho e^{i\varphi}$ where $\rho = e^{\pi x/a} > 0$ and $\varphi = \pi y/a$. In particular:

$$F(z_1 = 0) = w_1 = 1$$
$$F(z_2 = ia) = w_2 = -1 = 1e^{i\pi}$$

$$F(z_n = na/\pi + 0i) = w_n = e^n + 0i = e^n e^{i0} \qquad n > 0$$
$$F(z_m = ma/\pi + ai) = w_m = -e^m = e^m e^{i\pi} \qquad m > 0$$

$$F(z_p = -pa/\pi + 0i) = w_p = e^{-p} \qquad p > 0$$
$$F(z_q = -qa/\pi + ai) = w_q = e^{-q} e^{i\pi} = -e^{-q} \qquad q > 0$$

Note also that for fixed x_0, the vertical segment $\{z = x_0 + iy:\ 0 < y < a\}$ is mapped by F onto the semicircular arc $\{w = \rho e^{i\varphi}:\ \rho = e^{\pi x_0/a},\ 0 < \varphi < \pi\}$. This mapping is conformal at every point of the z-plane.

PROBLEM 6.9

Show that the mapping $w = \log z$ takes the region $\Omega = \{z = re^{i\vartheta}: r > 1, 0 < \vartheta < \pi/2\}$ into the semi-infinite strip $\Omega' = \{w = (u, v): u > 0, 0 < v < \pi/2\}$.

SOLUTION 6.9

The mapping function $F(z) = u + iv = \ln r + i\vartheta$, maps the z-plane with the branch cut along the positive real axis conformally onto the strip $\{w: -\infty < \operatorname{Re} w < \infty, 0 < \operatorname{Im} w < 2\pi\}$. The region Ω is pictured in Figure 6.10.

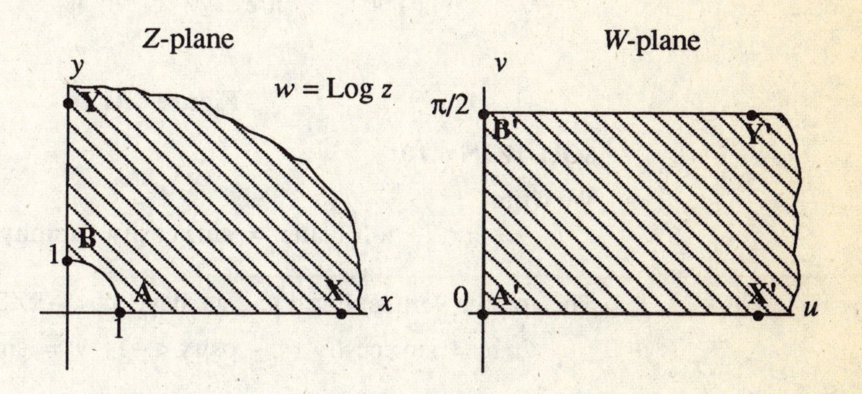

Figure 6.10

It is clear that the arc $AB = \{z = re^{i\vartheta}: r = 1, 0 < \vartheta < \pi/2\}$ is mapped by F onto the segment $A'B' = \{w = u + iv: u = \ln 1 = 0, 0 < v = \vartheta < \pi/2\}$. Similarly the rays $AX = \{z = re^{i\vartheta}: r = x \geq 1, \vartheta = 0\}$ and $BY = \{z = re^{i\vartheta}: r = y \geq 1, \vartheta = \pi/2\}$ are mapped onto $A'X' = \{u = \ln x \geq 0, v = \vartheta = 0\}$ and $B'Y' = \{u = \ln y \geq 0, v = \vartheta = \pi/2\}$. It is also evident that circular arcs in the z-plane of the form $\{z = Re^{i\vartheta}: 0 < \vartheta < \pi/2\}$ where $R > 1$ is fixed, are mapped onto vertical segments in the w-plane of the form $\{w = u + iv: u = \ln R > 0, 0 < v < \pi/2\}$. Thus the region Ω is mapped onto the strip Ω'.

PROBLEM 6.10

Show that the mapping function $F(z) = \sin z$ maps:

(a) the vertical strip shown in Figure 6.11 onto the upper half of the w-plane

(b) the complete vertical strip $|\operatorname{Re} z| < \pi/2$, onto the whole w-plane with two slits removed

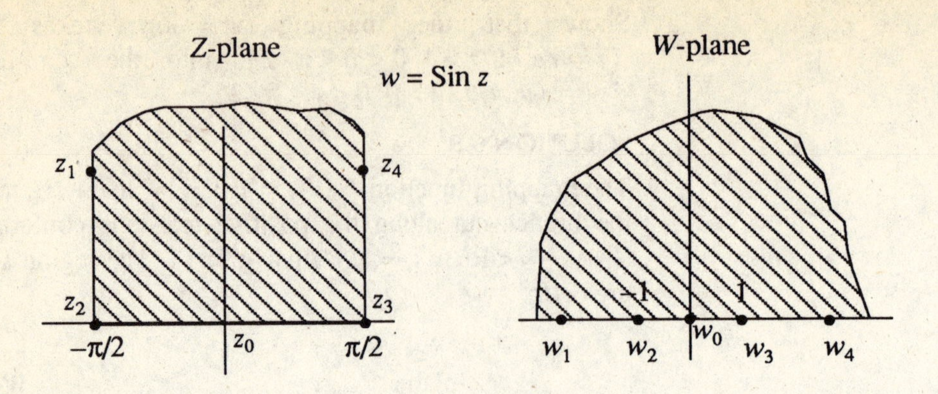

Figure 6.11

SOLUTION 6.10

We write

$$w = \sin z = \sin x \cosh y + i \sinh y \cos x$$

Then on the vertical line $z_1 z_2$, we have $x = -\pi/2$, $y > 0$, and

$$u = \sin x \cosh y = -\cosh y < -1, \quad v = \sinh y \cos x = 0.$$

Similarly, on $z_3 z_4$, $x = \pi/2$, $y > 0$, and

$$u = \sin x \cosh y = \cosh y > 1, \quad v = \sinh y \cos x = 0.$$

Thus the vertical sides of the strip fold down onto the segments $u < -1$ and $u > 1$ on the real axis in the w-plane. On the segment $z_2 z_3$ we have $y = 0$ and $|x| < \pi/2$. Then $|u| = |\sin x| < 1$ and $v = 0$ so the base of the strip is mapped to the interval $(-1, 1)$ on the real axis of the w-plane. Finally, note that on vertical lines $x = a$ inside the strip we have

$$u = \sin a \cosh y \quad \text{and} \quad v = \cos a \sinh y$$

Thus

$$\left(\frac{u}{\sin a}\right)^2 - \left(\frac{v}{\cos a}\right)^2 = 1$$

which implies that the vertical lines are mapped onto hyperbolic arcs in the upper half of the w-plane. Similarly, on horizontal lines $y = b$ inside the strip,

$$u = \sin x \cosh b \quad \text{and} \quad v = \cos x \sinh b$$

hence

$$\left(\frac{u}{\cosh b}\right)^2 + \left(\frac{v}{\sinh b}\right)^2 = 1$$

Then the horizontal lines in the strip map onto ellipses in the w-plane. The ellipses and hyperbolas in the w-plane are orthogonal families just as the horizontal and vertical lines are orthogonal in the z-plane.

The same arguments show that the lower half of the complete infinite strip $\{(x, y): -\pi/2 < x < \pi/2, \text{ all } y\}$ maps onto the lower half of the w-plane with the vertical sides of the half-strip going onto the segments $u < -1$ and $u > 1$. Then the mapping $w = F(z)$ is two to one for w on these two segments of the real axis of the w-plane. However the mapping is one to one from the inside of the complete strip onto the w-plane with slits removed along the two segments $\operatorname{Re} w < -1$ and $\operatorname{Re} w > 1$.

PROBLEM 6.11

Show that the mapping function $F(z) = z + 1/z$:

(a) maps the exterior of the unit disc in the z-plane onto the entire w-plane with a slit along the u-axis from –2 to 2

(b) maps a circle of the form $|z - z_0| = R$ onto an airfoil shaped region in the w-plane

SOLUTION 6.11

For the mapping $F(z) = z + 1/z$ we have $F'(z) = 1 - z^{-2}$. Then $F'(\pm 1) = 0$ and F is conformal at all z except $z = 1, -1$. If we write

$$w = u + iv = re^{i\vartheta} + \frac{1}{r}e^{-i\vartheta} = (r + 1/r)\cos\vartheta + i(r - 1/r)\sin\vartheta$$

then

$$u = (r + 1/r)\cos\vartheta \quad \text{and} \quad v = (r - 1/r)\sin\vartheta \tag{1}$$

For z on the unit circle, $|z| = r = 1$ and $u = 2\cos\vartheta$, $v = 0$. Hence the unit circle maps onto the interval $\{(u, v): -2 < u < 2, v = 0\}$ in the w-plane. For $r = R > 1$ we have

$$\frac{u^2}{(R + 1/R)^2} + \frac{v^2}{(R - 1/R)^2} = \cos^2\vartheta + \sin^2\vartheta = 1$$

Thus points on circles of radius $R > 1$ in the z-plane are mapped to ellipses in the w-plane. This proves (a).

We can write (1) in Cartesian form

$$u = x + \frac{x}{x^2 + y^2} \quad \text{and} \quad v = y - \frac{y}{x^2 + y^2} \tag{2}$$

Then the points (x, y) on circles of the form $(x - x_0)^2 + (y - y_0)^2 = R^2$ for given values of x_0, y_0 and R are mapped onto points (u, v) for $u(x, y)$ and $v(x, y)$ given by (2). The image of such circles is not easy to study analytically. We can show that if the

$y_0 = 0$ (so that the circle is symmetric with respect to the real axis), then the image of the circle in the w-plane will be symmetrical with respect to the u-axis. When the circle has no such symmetry in the z-plane it is not clear how the w-plane image will behave. Therefore we use a very simple computer program to plot the images for a selection of circles. The images are shown in Figure 6.12. In Figure 6.12(a) we see the image for a circle with $x_0 = 0.1$, $y_0 = 0$ and $R = 1$, and the image is symmetric about the u-axis as expected. Figure 6.12(b) shows the much blunter (but of course still symmetric), image of the circle with $x_0 = 0.4$, $y_0 = 0$. Figures 6.12(c) and (d) show the images of circles that are not symmetric with respect to the real axis in the z-plane. The images are "airfoil shapes" in the w-plane whose "camber" increases as the center of the circle in the z-plane moves away from the real axis.

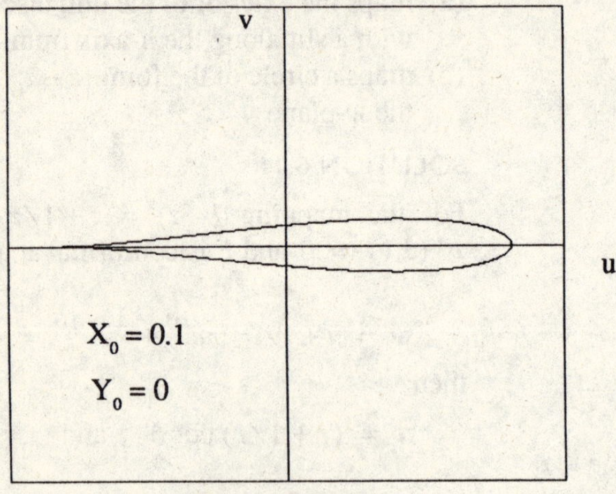

Figure 6.12(a)

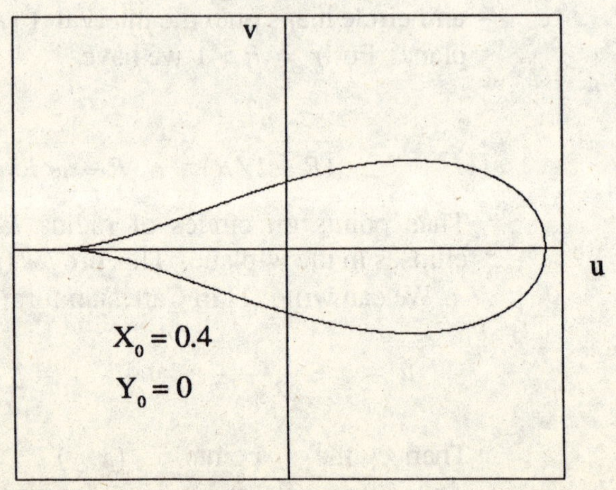

Figure 6.12(b)

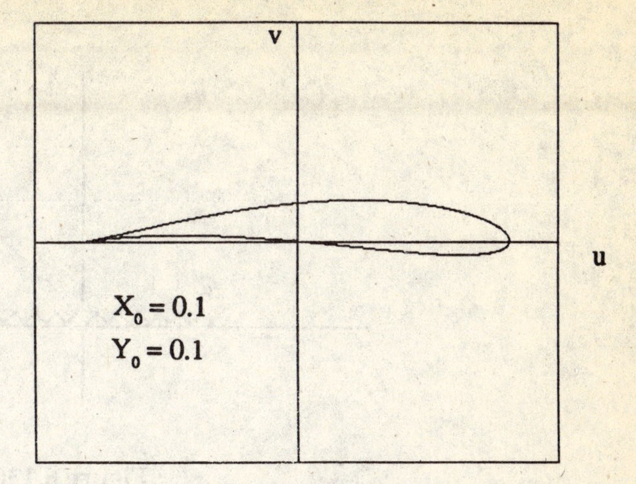

Figure 6.12(c)

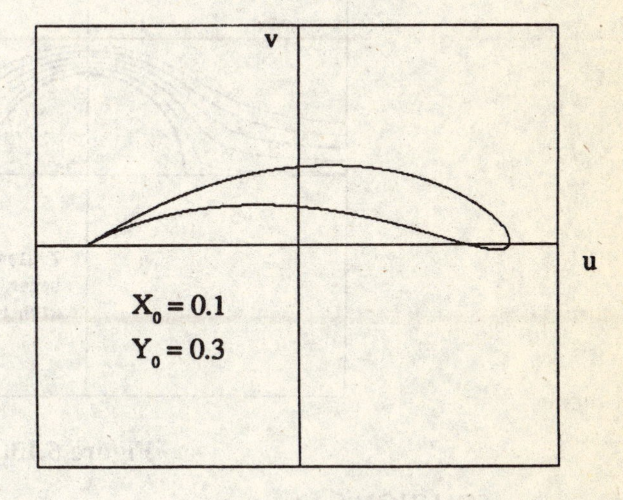

Figure 6.12(d)

PROBLEM 6.12

Show that the mapping function $F(z) = (z + \sqrt{z^2 - 4})/2$ takes the slit z-plane shown in Figure 6.13 onto the upper half of the w-plane with a unit half disc removed. Find the image in the w-plane of horizontal straight lines $y =$ Constant in the z-plane.

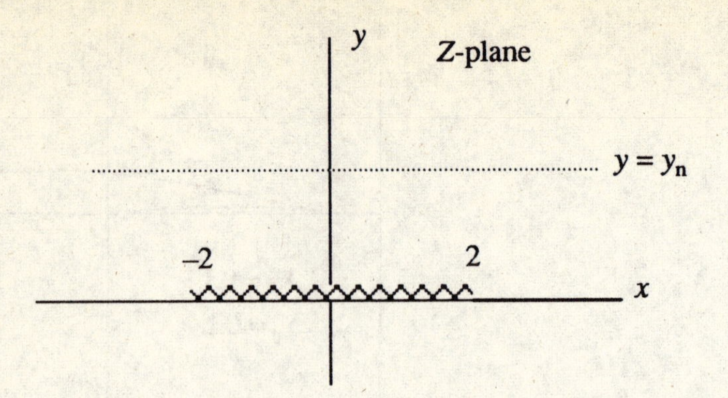

Figure 6.13(a)

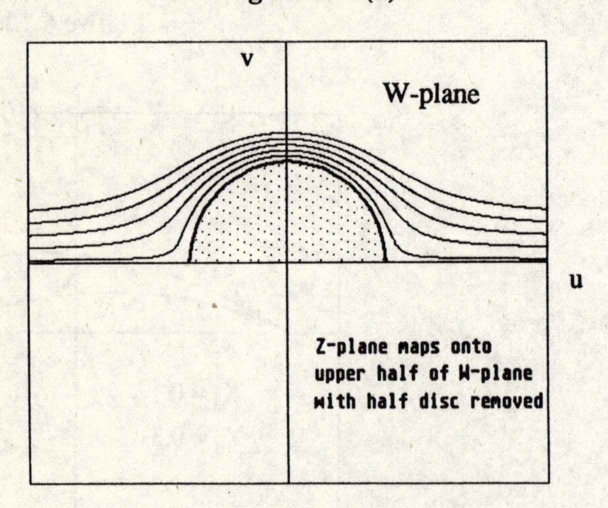

Figure 6.13(b)

SOLUTION 6.12

Solving the equation $2z = w + 1/w$, for $w = w(z)$ we find $w = (z + \sqrt{z^2 - 4})/2$. It follows that, except for a multiplicative factor of 2, this mapping $F(z)$ is the inverse of the mapping from the previous problem for z in the upper half plane. In particular, we can show that $F(z)$ has a branch cut on the real axis, extending from $z = -2$ to $z = 2$. $F(z)$ is not analytic at points on the branch cut but is analytic at all other points of the z-plane. We now proceed to find the image in the w-plane of the set consisting of the z-plane with the slit along the branch cut removed.

Writing

$$z = x + iy = \frac{1}{2}(w + \frac{1}{w}) = \frac{1}{2}(u + \frac{u}{u^2+v^2}) + i\frac{1}{2}(v - \frac{v}{u^2+v^2})$$

we see that $y = y_n$ (constant) implies

$$v - \frac{v}{u^2+v^2} = 2y_n \quad \text{or} \quad u^2+v^2 = \frac{v}{v-2y_n} \tag{1}$$

The equation (1) implicitly defines a curve C' in the w-plane which is the image of the horizontal straight line $C = \{(x, y): y = y_n\}$ in the z-plane. Note that for $v > 2y_n$, as v decreases toward $2y_n$, (1) implies that u^2 tends to plus infinity. We conclude from this that C' has a horizontal asymptote at $v = 2y_n$. Also, it follows from (1) that

$$u^2 = \frac{v(1+2y_nv - v^2)}{v-2y_n} \tag{2}$$

which implies $u = 0$ at $v = v_0 = y_n + \sqrt{(y_n)^2+1}$ and $dv/du = 0$ at $u = 0$. The images of several straight lines $y = y_n > 0$ are shown in Figure 6.13 where they are seen as bell shaped curves with horizontal asymptotes at $v = 2y_n$ and intercepts at $v = v_0$ on the vertical axis.

Successive Transformations

PROBLEM 6.13

Find a mapping to transform the infinite wedge $\Omega = \{z = re^{i\vartheta} : r > 0, 0 < \vartheta < \pi/4\}$ onto $\{|w| < 1\}$, the interior of the unit circle in the w-plane.

SOLUTION 6.13

We introduce the intermediate variable Z and note that the mapping $Z = F(z) = z^4$ maps Ω onto the upper half of the Z-plane. Now Problem 6.3 shows that the linear fractional transformation

$$w = \frac{Z-i}{Z+i} = G(Z)$$

maps the upper half of the Z-plane onto the interior of the unit circle in the w-plane. Then

$$w = \frac{z^4-i}{z^4+i} = G(F(z))$$

maps Ω into the interior of the unit circle.

PROBLEM 6.14

Find a mapping to transform the sector $\Omega = \{z = re^{i\vartheta}: 0 < r < R, 0 < \vartheta < \pi/3\}$ to the upper half of the w-plane.

SOLUTION 6.14

First, the mapping $Z_1 = F(z) = (z/R)^3$ maps the sector Ω onto the half-disc $D = \{Z = \rho e^{i\varphi}: 0 < \rho < 1, 0 < \varphi < \pi\}$. Next, the linear fractional transformation

$$Z_2 = -i\frac{Z_1 - 1}{Z_1 + 1} = G(Z_1)$$

sends the points $Z_1 = 1, i, -1$ on the unit circle in the Z_1-plane to the points $Z_2 = 0, 1, \infty$, respectively (see Problem 6.4). Moving around the unit circle $|Z_1| = 1$ in the order of the three points $Z_1 = 1, i, -1$, the interior of the circle is on the left. Visiting the image points $Z_2 = 0, 1, \infty$ in order in the Z_2-plane, the region on the left is the upper half of the Z_2-plane. Thus $Z_2 = G(Z_1)$ can be seen to map the interior of the unit disc in the Z_1 plane to the upper half of the Z_2-plane. Then the upper half of the disc can be seen to map onto the quarter plane, $\{Z_2 = X_2 + iY_2 : X_2 > 0, Y_2 > 0\}$. Finally, the mapping function

$$w = H(Z_2) = (Z_2)^2$$

maps the first quadrant of the Z_2-plane onto the upper half of the w-plane, (see Example 6.2(a)). Thus the compound mapping

$$w = \left(-i\frac{Z_1 - 1}{Z_1 + 1}\right)^2 = -\left(\frac{z^3 - R^3}{z^3 + R^3}\right)^2 = H(G(F(z)))$$

maps the sector onto the half-plane.

PROBLEM 6.15

Show that any linear fractional transformation of the form

$$w = e^{ic}\frac{z - z_0}{zz_0{}^* - 1} \qquad c = \text{real}, \ |z_0| < 1 \tag{1}$$

maps $|z| < 1$ onto $|w| < 1$.

SOLUTION 6.15

We are going to find successive linear fractional transformations that first carry $|z| < 1$ onto the upper half of the Z plane and then map $\operatorname{Im} Z > 0$ onto $|w| < 1$. If $Z = F(z)$ maps $|z| < 1$ onto the upper half plane $\operatorname{Im} Z > 0$, then $z = F^{-1}(Z)$ maps the upper half of the Z-plane onto $|z| < 1$. By the result of Problem 6.3, this mapping is of the form

$$z = \frac{Z - Z_0}{Z - Z_0{}^*} = F^{-1}(Z) \qquad \text{for } \operatorname{Im} Z_0 > 0 \tag{2}$$

and

$$Z = \frac{zZ_0^* - Z_0}{z - 1} \tag{3}$$

Now

$$w = \frac{Z - Z_1}{Z - Z_1^*} \quad \text{for } \operatorname{Im} Z_1 > 0 \tag{4}$$

maps $\operatorname{Im} Z > 0$ onto $|w| < 1$. Substituting (3) into (4)

$$w = \frac{\dfrac{zZ_0^* - Z_0}{z - 1} - Z_1}{\dfrac{zZ_0^* - Z_0}{z - 1} - Z_1^*} = \frac{Z_0^* - Z_1}{Z_0 - Z_1^*} \, \frac{z - z_0}{zz_0^* - 1} \tag{5}$$

where

$$z_0 = \frac{Z_0 - Z_1}{Z_0^* - Z_1}$$

Note that $|Z_0 - Z_1| < |Z_0^* - Z_1| = |Z_0 - Z_1^*|$ and thus $|z_0| < 1$ and

$$\left| \frac{Z_0^* - Z_1}{Z_0 - Z_1^*} \right| = 1 \quad \text{i.e.,} \quad \frac{Z_0^* - Z_1}{Z_0 - Z_1^*} = e^{ic} \quad \text{for } c = \text{real}$$

Then (5) has the form (1) and maps $|z| < 1$ onto $|w| < 1$.

Miscellaneous Mappings

PROBLEM 6.16

Find a mapping that takes the shaded area in Figure 6.14(a) onto the annular domain $\{w\colon\ R < |w| < 1\}$ shown in Figure 6.14(b).

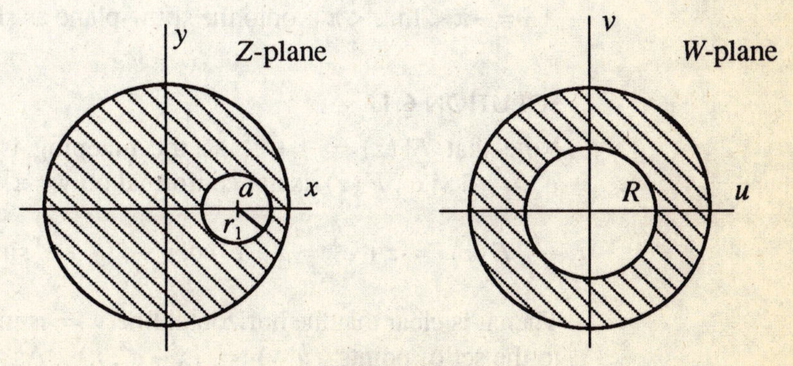

Figure 6.14

SOLUTION 6.16

The shaded area in Figure 6.14(a) lies inside the disc $\{|z| < 1\}$ but outside the smaller disc, $\{|z-a| < r_1\}$. Note that a and r_1 are given with $r_1 + a < 1$ and $a - r_1 > 0$. By the result of the previous problem, the mapping

$$w = \frac{z - z_0}{z z_0^* - 1} = F(z) \tag{1}$$

maps $|z| < 1$ onto $|w| < 1$ for any z_0 with $|z_0| < 1$. We will choose $z_0 = C$ for $C =$ real where C is to be chosen such that $z_1 = a - r_1$ maps to $w_1 = -R > -1$, and the point $z_2 = a + r_1$ maps to $w_2 = R$. That is, C must be such that

$$F(z_1) = \frac{a - r_1 - C}{(a - r_1)C - 1} = \frac{a + r_1 - C}{(a + r_1)C - 1} = -F(z_2)$$

i.e.,

$$aC^2 - (1 + a^2 - (r_1)^2)C + a = 0.$$

This quadratic equation for C has two real roots. The root such that $0 < C < 1$ is given by

$$C = (1 + a^2 - (r_1)^2 - \sqrt{(1 + a^2 - (r_1)^2)^2 - 4a^2})/2a \tag{2}$$

and

$$R = F(z_1) = \frac{a - r_1 - C}{(a - r_1)C - 1} \tag{3}$$

for C given by (2). Then the desired mapping onto the annulus is given by (1) for $z_0 = z_0^* = C$.

PROBLEM 6.17

Show that the mapping $F(z) = z + e^z$ takes the z-plane strip $\{S = -\pi < \operatorname{Im} z < \pi\}$ onto the slit w-plane as shown in Figure 6.15.

SOLUTION 6.17

Note that $F'(z) = 1 + e^z$ so the mapping is conformal except where $e^z = -1$; i.e., $F(z)$ is not conformal on $y = \pi, -\pi$. We write

$$F(z) = z + e^z = x + e^x \cos y + i(y + e^x \sin y) = u + iv \tag{1}$$

Then it is clear that the horizontal line $y = \pi$ in the z-plane is transformed to the set of points $(u, v) = (x - e^x, \pi)$. As x runs through all real values from minus infinity up to zero, the expression $u = x - e^x$ increases

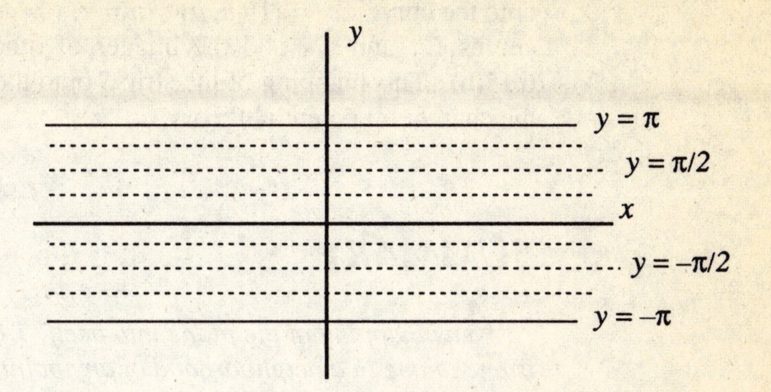

Figure 6.15(a)

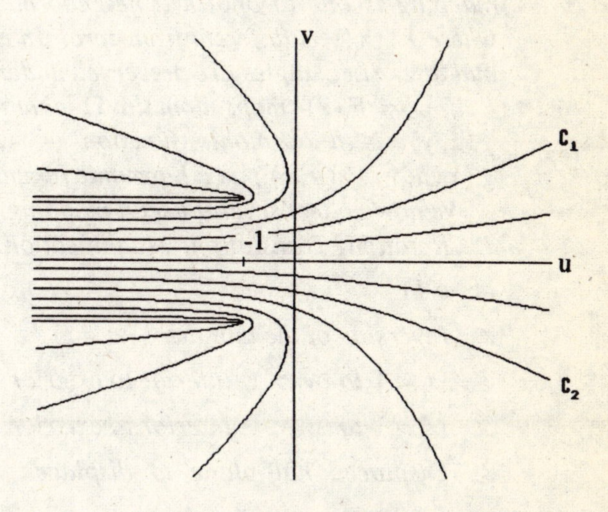

Figure 6.15(b)

from minus infinity to -1. As x runs through all real values between zero and plus infinity, $u = x - e^x$ goes from -1 down to minus infinity. Thus the horizontal line is mapped onto the semi-infinite line $\{(u, v\!: u \le -1, v = \pi\}$, with each point on the line corresponding to two points on $y = \pi$. Similarly, the line $y = -\pi$ maps in the same two to one fashion onto the semi-infinite line $\{(u, v): u \le -1, v = -\pi\}$. The inside of the strip S is mapped conformally onto the w-plane from which the two semi-infinite strips have been removed.

The x-axis in the z-plane maps conformally onto the u-axis in the w-plane. Similarly, horizontal lines inside the strip S are mapped conformally onto curves in the w-plane. For example the line $\{(u, v): y = \pi/2\}$ is mapped onto the curve$C_1 = \{(u, v\!: u = x,$

$v = e^x + \pi/2, -\infty < x < \infty\}$ and the line $\{(u, v): y = -\pi/2\}$ is mapped onto the curve $C_2 = \{(u, v): u = x, v = -e^x - \pi/2, -\infty < x < \infty\}$. The curves C_1 and C_2 and the images of other lines are shown in Figure 6.15(b). This mapping of the strip S onto the slit w-plane will be of use in the chapter on potential theory.

SUMMARY

Any mapping of the plane into itself, $T(x, y) = (u(x, y), v(x, y))$, *is one to one in a neighborhood of any point* (x_0, y_0) *where the Jacobian is different from zero. If the transformation is analytic, (i.e., if* $F(x + iy) = (u(x, y) + iv(x, y))$ *is analytic), then* $J(x_0, y_0) = |F'(z_0)|^2$ *hence the mapping is one to one in a neighborhood of any point* $z_0 = (x_0, y_0)$ *where* $F'(z_0)$ *is different from zero. In addition, the mapping is conformal at* z_0 *(i.e., angles are preserved under the mapping).*

If $w = F(z)$ *maps domain* Ω *conformally into domain* Ω' *and if* $g(x, y)$ *is a harmonic function of* (x, y) *in* Ω, *then* $G(u, v) = g(x(u, v), y(u, v))$ *is a harmonic function of* (u, v) *in* Ω'.

We now collect several useful mapping functions:

1. *Rotation, Translation, Magnification:* $w = Az + B$ *(See Example 6.1)*
2. *Inversion of the unit disc:* $w = 1/z$ *(See Example 6.3(a))*

   $|z| < 1$ *to* $|w| > 1$ *interior to exterior*

   $|z| > 1$ *to* $|w| < 1$ *interior to exterior*
3. *Displaced half-plane to displaced disc:* $w = 1/z$ *(See Example 6.3(b))*

   $y > D$ *to* $u^2 + (v + 1/2D)^2 < (1/2D)^2$

   $x > C$ *to* $(u - 1/2C)^2 + v^2 < (1/2C)^2$
4. *Displaced Disc to displaced half-plane:* $w = 1/z$ *(See Example 6.3(c))*

   $(x - a)^2 + y^2 < a^2$ *to* $u > 1/2a$

   $x^2 + (y + b)^2 < b^2$ *to* $v > 1/2b$
5. *Wedge to upper half-plane:* $w = z^p$ *where* $p = \pi/\alpha$ *(See Example 6.2, Problem 6.7)*

   $\{(r, \vartheta): r > 0, 0 < \vartheta < \alpha\}$ *to* $\operatorname{Im} w > 0$
6. *Horizontal Strip to upper half-plane:* $w = e^{\pi z/a}$ *(See Problem 6.8)*

$\{0 < \operatorname{Im} z < a\}$ *to* $\operatorname{Im} w > 0$

7. *Vertical Strip to upper half-plane:* $w = \sin z$ *(See Problem 6.10)*

   $\{(x, y): |x| < \pi/2, y > 0\}$ *to* $\operatorname{Im} w > 0$

8. *Upper half-plane to unit disc:* $w = \dfrac{z-i}{z+i}$ *(See Problem 6.3)*

   $\operatorname{Im} z > 0$ *to* $|w| < 1$

9. *Right half-plane to unit disc:* $w = \dfrac{z-1}{z+1}$ *(See Problem 6.3)*

   $\operatorname{Re} z > 0$ *to* $|w| < 1$

10. *Unit disc to unit disc:* $w = e^{ic}\dfrac{z-\upsilon}{z\upsilon^* - 1}, |\upsilon| < 1$ *(See Problem 6.17)*

    $|z| < 1$ *to* $|w| < 1$

11. *Right half-plane with disc removed to unit disc:* $w = i/(z-1)$ *(See Problem 6.5)*

    $\{(x, y): (x-1)^2 + y^2 > 1, x > 0\}$ *to* $|w| < 1$

12. *Right half-plane with disc removed to strip:* $w = 1/z$ *(See Problem 6.6)*

    $\{(x, y): (x-a)^2 + y^2 > a^2, x > 0\}$ *to* $0 < \operatorname{Re} w < 1/2a$

13. *Wedge with sector removed to strip:* $w = \log z$ *(See Problem 6.9)*

    $\{(r, \vartheta): r > 1, 0 < \vartheta < \pi/2\}$ *to* $\{(u, v): u > 0, 0 < v < \pi/2\}$

    or, $\{(r, \vartheta): r > 1, 0 < \vartheta < \pi\}$ *to* $\{(u, v): u > 0, 0 < v < \pi\}$

14. *Unit disc with circular hole to annulus:* $w = \dfrac{z-C}{Cz-1}$ *(See Problem 6.16)*

    $\{z: |z| < 1, |z-a| < r_1\}$ *to* $\{w: R < |w| < 1\}$

    $$C = (1 + a^2 - (r_1)^2 - \sqrt{(1 + a^2 - (r_1)^2)^2 - 4a^2})/2a$$

    $$R = F(z_1) = \frac{a - r_1 - C}{(a - r_1)C - 1}$$

7

Applications of Complex Integration

We have seen that the residue theorem for complex contour integrals can be used to evaluate certain improper real integrals that are not tractable by standard calculus techniques. Contour integration methods are particularly suited to evaluating improper integrals that arise in connection with integral transforms. In this chapter we apply contour integration in connection with the integral transforms of Fourier and Laplace as these are the transforms that are used most frequently in applications.

We have also seen in an earlier chapter that the argument principle for complex contour integrals can be used to locate the zeros of polynomials. We show here how this may be used in connection with the Laplace transform to determine the stability of linear systems.

INTEGRAL TRANSFORMS

For a function $f(x)$ defined on the interval (a, b) we define an *integral transform* of f to be a function $F(\lambda)$, given by an expression of the form

$$T\{f\}(\lambda) = \int_a^b K(x, \lambda) f(x)\, dx = F(\lambda)$$

The given function, $K(x, \lambda)$ is called the *kernel* of the integral transform. As we shall see, integral transforms are often useful in solving differential and integral equations. In this chapter we consider two integral transforms with kernels based on the exponential function, the transforms of Fourier and Laplace.

FOURIER TRANSFORM

For $f(x)$ defined on the whole real line, the Fourier transform of f is defined to be the function $F(\alpha)$ given by

$$T_F\{f\}(\alpha) = \frac{1}{2\pi}\int_{-\infty}^{\infty} e^{-ix\alpha} f(x)\,dx = F(\alpha) \tag{7.1}$$

We use the notations $T_F\{f\}$ and $F(\alpha)$ for the Fourier transform interchangeably.

SQUARE INTEGRABLE FUNCTIONS

The Fourier transform is defined in terms of an improper integral, hence the integral does not exist for all functions $f(x)$. There are various function classes on which the transform is well defined, one of which is the class of square integrable functions. The function $f(x)$ is said to be *square integrable* if

$$\int_{-\infty}^{\infty} |f(x)|^2 dx < \infty \tag{7.2}$$

Here $|f(x)|$ denotes the modulus of $f(x)$. Although we will be dealing primarily with real valued functions $f(x)$, we state the definition of square integrability so that it may be applied to functions that are complex valued as well.

THE FOURIER INVERSION THEOREM

Theorem 7.1

For each square integrable function $f(x)$, the Fourier transform $F(\alpha)$ exists and is a square integrable function of the transform variable, α. Moreover, each square integrable function F is the Fourier transform of a square integrable function f related to F by the Fourier inversion formula,

$$f(x) = \int_{-\infty}^{\infty} F(\alpha)\, e^{ix\alpha} d\alpha = T_F^{-1}\{F\} \tag{7.3}$$

EXAMPLE 7.1

(a) For positive real number c let

$$I_c(x) = \begin{cases} 1 & \text{if } |x| < c \\ 0 & \text{if } |x| > c \end{cases}$$

Clearly $I_c(x)$ is square integrable and

$$T_F(I_c) = (2\pi)^{-1}\int_{-\infty}^{\infty} I_c(x)\, e^{-ix\alpha} dx = (2\pi)^{-1}\int_{-c}^{c} e^{-ix\alpha} dx$$

$$= \frac{-1}{2\pi i\alpha} e^{-ix\alpha}\Big|_{-c}^{c} = \frac{\sin\alpha c}{\pi\alpha}$$

(b) The function $F(\alpha) = e^{-b|\alpha|}$ is a square integrable function of α for any positive real number, b. Then we may apply the Fourier inversion formula to find a square integrable function, $f(x)$, whose transform is F: i.e.,

$$T_F^{-1}\{e^{-b|\alpha|}\} = \int_{-\infty}^{\infty} e^{-b|\alpha|} e^{ix\alpha}\, d\alpha$$

$$= \int_{-\infty}^{0} e^{b\alpha} e^{ix\alpha}\, d\alpha + \int_{0}^{\infty} e^{-b\alpha} e^{ix\alpha}\, d\alpha$$

$$= \frac{e^{(b+ix)\alpha}}{b+ix}\Big|_{-\infty}^{0} + \frac{e^{(-b+ix)\alpha}}{-b+ix}\Big|_{0}^{\infty} = \frac{1}{b+ix} - \frac{1}{-b+ix}$$

$$= \frac{2b}{b^2+x^2}$$

(c) For positive real number c, $F(\alpha) = -c\alpha^2$ is square integrable. Then the inversion formula may be applied to this F to find $f = T_F^{-1}\{F\}$,

$$T_F^{-1}\{e^{-c\alpha^2}\} = \int_{-\infty}^{\infty} e^{-c\alpha^2} e^{ix\alpha}\, d\alpha$$

$$= \int_{-\infty}^{\infty} e^{-c(\alpha^2 - ix\alpha/c - x^2/4c^2 + x^2/4c^2)}\, d\alpha$$

$$= e^{-x^2/4c} \int_{-\infty}^{\infty} e^{-c(\alpha - ix/2c)^2}\, d\alpha$$

$$= e^{-x^2/4c} \int_{-\infty}^{\infty} e^{-c\beta^2}\, d\beta = e^{-x^2/4c}\sqrt{\pi/c}$$

FOURIER TRANSFORM PAIRS

The function pairs:

$I_c(x)$	$\frac{\sin\alpha c}{\pi\alpha}$	$c > 0$
$\frac{2b}{b^2+x^2}$	$e^{-b\|\alpha\|}$	$b > 0$
$e^{-x^2/4c}\sqrt{\pi/c}$	$e^{-c\alpha^2}$	$c > 0$

are examples of *Fourier transform pairs,* $(f(x), F(\alpha))$; i.e., $F = T_F\{f\}$ in each case.

RIEMANN AND LEBESGUE INTEGRALS

The proof of Theorem 7.1 relies on results from the theory of Lebesgue integration which means that the integrals in (7.1) and (7.3) are understood to be Lebesgue integrals. In this chapter, however, we will be computing Fourier transforms for Lebesgue integrable functions that are also Riemann integrable. Since the two integrals agree in value for functions that are both Riemann integrable and Lebesgue integrable, we may consider all integrals in this chapter to be Riemann integrals. In fact, we will often evaluate integrals of the form (7.1) and (7.3) by replacing them by suitable complex contour integrals which will be evaluated by means of residues. We have the following restatement of the result of Problem 5.14 for this purpose. This is a version of what is usually referred to as *Jordan's lemma.*

Theorem 7.2

Let Γ_R denote the closed contour consisting of the interval $(-R, R)$ on the real axis, followed by the semicircular arc $C_R = \{(r, \vartheta)\colon\ r = R, 0 < \vartheta < \pi\}$. Suppose $F(\alpha)$ satisfies:

(a) $F(z)$ is analytic except for finitely many isolated singularities $z_1, \ldots, z_m$ in the upper half-plane
(b) for each $\varepsilon > 0$ there exists an $R > R_0$ such that $\left|e^{ixz}F(z)\right| < \varepsilon$ for z on C_R

where R_0 denotes the largest of the numbers $|z_1|, \ldots, |z_m|$. Then

$$\int_{-\infty}^{\infty} F(\alpha)\, e^{ix\alpha} dx = \lim_{R\to\infty} \int_{\Gamma_R} F(z)\, e^{ixz} dz = 2\pi i \sum_{j=1}^{m} \operatorname{Res} F(z_j)\, e^{ixz_j}$$

The theorem is also true when F has singularities in the lower half-plane and the semicircular arc C_R is then placed in the lower half-plane as well.

EXAMPLE 7.2

Consider the function $F(\alpha) = 1/(b^2 + \alpha^2)$ for $b > 0$. It is easy to check that F is square integrable hence

$$T_F^{\ -1}\{F\} = \int_{-\infty}^{\infty} \frac{e^{ix\alpha}}{b^2 + \alpha^2} d\alpha$$

For $x > 0$, conditions (a) and (b) of Theorem 7.2 are satisfied by the integrand $e^{ixz}(b^2 + z^2)^{-1}$. Since $F(z) = 1/(b^2 + \alpha^2)$ has only a simple pole at $z = ib$ in the upper half-plane, we find

$$T_F^{-1}\{\frac{1}{b^2+\alpha^2}\} = 2\pi i\frac{e^{-xb}}{2ib} = \frac{\pi}{b}e^{-xb} \quad x>0.$$

For $x<0$, condition (b) is not satisfied. However, on the closed contour $\Gamma_R{}'$ consisting of the straight line from $(R,0)$ to $(-R,0)$ followed by the semicircle $C_R{}' = \{(r,\vartheta) : r=R, \pi<\vartheta<2\pi\}$ the analogue of condition (b) does hold. The contour $\Gamma_R{}'$ encloses the simple pole at $z = -ib$ and thus for $x<0$

$$\int_{-\infty}^{\infty} F(\alpha)\,e^{ix\alpha}dx = -\lim_{R\to\infty}\int_{\Gamma_R{}'} F(z)\,e^{ixz}dz$$

$$= -2\pi i\frac{e^{xb}}{-2ib} = \frac{\pi}{b}e^{xb} \quad x<0.$$

Therefore

$$T_F^{-1}\{\frac{1}{b^2+\alpha^2}\} = \begin{cases} \frac{\pi}{b}e^{xb} & x<0 \\ \frac{\pi}{b}e^{-xb} & x>0 \end{cases} = \frac{\pi}{b}e^{-b|x|}$$

i.e. the functions

$$\pi/be^{-b|x|} \qquad \frac{1}{b^2+\alpha^2}$$

are a Fourier transform pair.

Operational Properties of the Fourier Transform

The following properties hold for arbitrary square integrable functions f and g and their Fourier transforms. These properties are referred to as *operational properties* of the Fourier transform and are the basis for the usefulness of the transform in solving problems in differential and integral equations.

Theorem 7.3

Let f and g be arbitrary square integrable functions with Fourier transforms F and G respectively. Then

1. For arbitrary constants a and b: $T_F\{af+bg\} = aF(\alpha)+bG(\alpha)$

2. For nonzero constant a, $T_F\{f(ax)\} = \frac{1}{|a|}F(\alpha/a)$

3. If f and f' are both square integrable then $T_F\{f'\} = i\alpha F(\alpha)$
   If f, f' and f'' are all square integrable then $T_F\{f''\} = -\alpha^2 F(\alpha)$

4. If $xf(x)$ is square integrable then $T_F\{xf(x)\} = iF'(\alpha)$

5. For any real constant c, $T_F\{f(x-c)\} = e^{-i\alpha c}F(\alpha)$

6. For any real constant c, $T_F\{e^{icx}f(x)\} = F(\alpha - c)$

7. If f is square integrable then $F(\alpha)$ is square integrable as well and

$$T_F\{F(x)\} = f(-\alpha)/2\pi$$

8. If we define the convolution product, $f*g$, of f and g by

$$f*g(x) = \int_{-\infty}^{\infty} f(x-y)\,g(y)\,dy$$

then we have $T_F\{f*g(x)\} = 2\pi F(\alpha)G(\alpha)$; i.e., $T_F{}^{-1}\{FG\} = f*g/2\pi$.

The properties 3 and 8 are the properties most often used in solving differential equations.

EXAMPLE 7.3

Consider the problem of finding unknown square integrable function $y = y(x)$ which has square integrable first and second derivatives and solves the equation

$$y''(x) - b^2 y(x) = f(x) \quad \text{for} \quad -\infty < x < \infty$$

for arbitrary given square integrable function $f(x)$. If we let $Y(\alpha)$, $F(\alpha)$ denote the Fourier transforms of y and f then $T_F\{y''(x)\} = -\alpha^2 Y(\alpha)$ by property 3 of Theorem 7.3. Then it follows that

$$-(\alpha^2 + b^2)\,Y(\alpha) = F(\alpha)$$

i.e.,

$$Y(\alpha) = -F(\alpha)\,G(\alpha) \quad \text{for} \quad G(\alpha) = \frac{1}{\alpha^2 + b^2}$$

By Property 8, we have

$$\begin{aligned} y(x) &= \frac{-1}{2\pi}\int_{-\infty}^{\infty} g(x-\xi)\,f(\xi)\,d\xi \\ &= \frac{-1}{2b}\int_{-\infty}^{\infty} e^{-b|x-\xi|}f(\xi)\,d\xi \end{aligned}$$

where we have used the result of Example 7.2 to discover $T_F^{-1}\{G\}$.

THE LAPLACE TRANSFORM

For $f(t)$ defined on the half-line $[0, \infty)$ we define the Laplace transform of f to be the function $F(s)$ given by

$$T_L\{f(t)\} = \int_0^\infty e^{-st} f(t)\,dt = F(s) \tag{7.4}$$

The notations, $T_L\{f\}$ and $F(s)$, for the Laplace transform of f, are synonymous.

ADMISSIBLE FUNCTIONS

In order for the function $f(t)$ to have a Laplace transform $F(s)$, it is sufficient that the improper integral (7.4) exists. The integral is guaranteed to exist if:

(a) $f(t)$ has a finite number of finite jump discontinuities on any interval of finite length

(7.5)

(b) there exist constants M and b such that

$$|f(t)| \le Me^{bt} \quad \text{for all } t > 0$$

A function satisfying (7.5a) is said to be *piecewise continuous* and if f satisfies (7.5b), we say f is of *exponential type.* Every function satisfying both conditions has a Laplace transform but not every function with a Laplace transform necessarily satisfies these conditions.

Theorem 7.4

Suppose the real valued function $f(t)$ satisfies (7.5) for constants M and b. Then f has a Laplace transform $F(z)$, $z = x + iy$, which is an analytic function of z for $\operatorname{Re} z > b$.

EXAMPLE 7.4

The function $f(t) = 1$ satisfies the conditions (7.5) and therefore has a Laplace transform

$$T_L\{1\} = \int_0^\infty e^{-st}dt = \left.\frac{e^{-st}}{-s}\right|_0^\infty = \frac{1}{s} \quad \text{for } s > 0.$$

We say that $f(t) = 1$ and $F(s) = 1/s$ form a *Laplace transform pair.* Note that the function that is identically 1 is not square integrable and thus does not have a square integrable Fourier transform.

THE LAPLACE TRANSFORM INVERSION FORMULA

Recovery of the function $f(t)$ from its Laplace transform $F(s)$ is accomplished by taking advantage of the fact that $F(z)$ is analytic for $\operatorname{Re} z > b$ and using a version of the Cauchy integral formula.

Theorem 7.5

Suppose $F(z)$ is a complex valued function of the complex variable z such that $F(x)$ is real valued for x real. Suppose also that $F(z)$ is analytic in the half-plane $\operatorname{Re} z > b$ and that for constants $M > 0$ and $m > 1$, $|z^m F(z)| < M$ for all z, $\operatorname{Re} z > b$. Then for any $c > b$

$$f(t) = \frac{1}{2\pi i} \lim_{p \to \infty} \int_{c-ip}^{c+ip} e^{zt} F(z)\, dz \tag{7.6}$$

Here $f(t)$ is independent of c and $L_T\{f\} = F$. Moreover, f satisfies (7.5) and $f(t) = 0$ for $t < 0$.

The condition that $|z^m F(z)| < M$ for $\operatorname{Re} z > b$ is sufficient but not necessary for $F(z)$ to be the Laplace transform of some $f(t)$.

PROPERTIES OF THE LAPLACE TRANSFORM

The Laplace transform has properties similar to the operational properties of the Fourier transform which are similarly useful in solving differential and integral equations.

Theorem 7.6 Operational Properties of the Laplace Transform

Let $f(t)$, $g(t)$ denote arbitrary functions with $T_L\{f(t)\} = F(s)$ and $T_L\{g(t)\} = G(s)$. Then

1. $T_L\{af(t) + bg(t)\} = aF(s) + bG(s)$ for all constants a, b
2. $T_L\{f(at)\} = (1/a) F(s/a)$ for a not equal to zero.
3. $T_L\{f'(t)\} = sF(s) - f(0)$
   $T_L\{f''(t)\} = s^2 F(s) - sf(0) - f'(0)$
   $T_L\{f^{(n)}(t)\} = s^n F(s) - s^{n-1} f(0) - \ldots - f^{(n-1)}(0)$
4. $T_L\{tf(t)\} = -F'(s)$
5. $T_L\{e^{bt} f(t)\} = F(s-b)$ for all real constants b
6. $T_L\{H(t-b) f(t-b)\} = e^{-bs} F(s)$ for all real constants b

where $H(t) = \textit{Heaviside step function} = \begin{cases} 0 & \text{if } t \le 0 \\ 1 & \text{if } t > 0 \end{cases}$

7. $T_L\{f*g(t)\} = F(s)G(s)$

where $f*g(t) = \int_0^t f(t-T)g(T)\,dT$

THE CONVOLUTION PRODUCT

The product $f*g(t)$ defined in part 7 of Theorem 7.5 is called the *convolution product* of the functions f and g. The convolution product has the following properties:

$$f*g(t) = \int_0^t f(t-T)g(T)\,dT = \int_0^t f(T)g(t-T)\,dT = g*f(t)$$

$$f*(g+h)(t) = f*g(t) + f*h(t)$$

STABILITY OF LINEAR SYSTEMS

Consider a linear system where the *output*, $u(t)$, is related to the *input*, $f(t)$, by an initial value problem of the form

$$u^{(n)}(t) + a_{n-1}u^{(n-1)}(t) + \ldots + a_1u'(t) + a_0 = f(t)$$

$$u(0) = u'(0) = \ldots = u^{(n-1)}(0) = 0.$$

Here $a_0, \ldots, a_{n-1}$ denote real constants. Then $U(s)$, the Laplace transform of the output, is related to $F(s) = T_F\{f\}$ by $U(s) = G(s)F(s)$ where

$$G(s) = \frac{1}{P(s)} = \frac{1}{s^n + a_{n-1}s^{n-1} + \ldots + a_1s + a_0}$$

THE TRANSFER FUNCTION

We refer to $G(s)$ as the *transfer function* for the linear system. Then part 7 of Theorem 7.6 implies that the output is given by

$$u(t) = \int_0^t g(t-\tau)f(\tau)\,d\tau = \int_0^t g(\tau)f(t-\tau)\,d\tau$$

where $g = T_L^{-1}\{G\}$. A few examples of G and its inverse are as follows:

$G(s)$	$g(t)$
$\dfrac{1}{s-a}$	e^{at}
$\dfrac{s}{s^2+b^2}$	$\cos bt$
$\dfrac{s-a}{(s-a)^2+b^2}$	$e^{at}\cos bt$
$\dfrac{2as}{(s^2+a^2)^2}$	$t\sin at$

STABLE LINEAR SYSTEMS

A function $f(t)$ is said to be *bounded* if there exists a positive constant M such that $|f(t)| \le M$ for all $t \ge 0$. Then a linear system will be said to be *stable* if every bounded input $f(t)$, produces a bounded output $u(t)$. More precise definitions of stability for linear systems are possible but will not be considered here.

Theorem 7.7 The linear system whose transfer function is $G(s) = 1/P(s)$ is stable if and only if:

1. $P(s)$ has no zeros in the right half of the complex plane.
2. Any zeros of $P(s)$ which occur on the imaginary axis are simple zeros.

EXAMPLE 7.5

(a) Theorem 7.7 implies that the system whose transfer function is $G(s) = b/(s^2+b^2)$ is stable since $P(s) = s^2+b^2$ has only the simple zeros $z = \pm ib$. In fact if $|f(t)| \le M$ for all $t \ge 0$ then we can see directly

$$|u(t)| = \left|\int_0^t f(t-\tau)\cos b\tau\, d\tau\right| \le M\left|\int_0^t \cos b\tau\, d\tau\right| \le \frac{M}{b};$$

i.e., the bounded input $f(t)$ produces a bounded output.

(b) For $G(s) = 1/(s-2)$, the bounded input $f(t) = 1$ produces the unbounded output

$$u(t) = \int_0^t f(t-\tau)\, e^{2\tau} d\tau = \frac{1}{2}(e^{2t} - 1)$$

This instability could have been anticipated from Theorem 7.7 since $P(s) = s-2$ has a zero in the right half plane.

(c) For $G(s) = (2s)/(s^2+1)^2$, $P(s) = (s^2+1)^2$ has double zeros at $z = \pm i$ on the imaginary axis and Theorem 7.7 implies the system whose transfer function is $G(s)$ is unstable. This is illustrated by the fact that the bounded input $f(t) = 1$ produces the unbounded output

$$u(t) = \int_0^t \tau \sin \tau\, d\tau = \sin t - t \cos t.$$

SOLVED PROBLEMS

Fourier Transform

PROBLEM 7.1

Use the Fourier inversion formula to find $T_F^{-1}\{\frac{2\sin\alpha}{\alpha}\}$.

SOLUTION 7.1

According to the Fourier inversion formula

$$T_F^{-1}\{\frac{2\sin\alpha}{\alpha}\} = \int_{-\infty}^{\infty} \frac{2\sin\alpha}{\alpha} e^{ix\alpha} d\alpha$$

$$= \int_{-\infty}^{\infty} \frac{1}{i\alpha}(e^{i(1+x)\alpha} - e^{-i(1-x)\alpha})\, d\alpha$$

The integrand here has only a simple pole at the origin. If $x < -1$ then we choose a contour Γ, like the one pictured in Figure 5.17, running from $-R$ to $-\varepsilon$ along the negative real axis and then around the semicircle of radius ε in the upper half plane. The contour continues from ε to R along the positive real axis and then around a semicircular contour of radius R in the upper half plane. Then for $x < -1$, Jordan's lemma implies

$$\lim_{\substack{R\to\infty \\ \varepsilon\to 0}} \int_\Gamma \frac{1}{iz}(e^{i(1+x)z} - e^{-i(1-x)z})\, dz = \int_{-\infty}^{\infty} \frac{2\sin\alpha}{\alpha} e^{ix\alpha} d\alpha$$

But the result of Problem 5.16 implies that for all ε, $R > 0$ the contour integral equals

$$\int_{\Gamma_R} \frac{1}{iz}\left(e^{i(1+x)z} - e^{-i(1-x)z}\right) dz = \pi i \text{Res}\left(\left(e^{i(1+x)z} - e^{-i(1-x)z}\right)/i\right)\Big|_{z=0}$$
$$= 0.$$

For $-1 < x < 1$, we obtain in the same way

$$\lim_{\substack{R \to \infty \\ \varepsilon \to 0}} \left(\int_{\Gamma} \frac{1}{iz} e^{i(1+x)z} dz + \int_{\Gamma'} e^{-i(1-x)z} \frac{1}{iz} dz \right) = \int_{-\infty}^{\infty} \frac{2\sin\alpha}{\alpha} e^{ix\alpha} d\alpha$$

where Γ' denotes the contour Γ rotated 180 degrees about the origin. Note that then Γ' runs from R to $-R$ along the real axis which requires the sign change in the integrand of Γ' to obtain the equality (1). Then we have

$$\pi i \text{Res}\left(e^{i(1+x)z}/i\right)\Big|_{z=0} + \pi i \text{Res}\left(e^{-i(1-x)z}/i\right)\Big|_{z=0} = 2\pi$$

and

$$\int_{-\infty}^{\infty} \frac{2\sin\alpha}{\alpha} e^{ix\alpha} d\alpha = 2\pi \quad \text{for } -1 < x < 1.$$

Finally for $x > 1$ we integrate around the contour Γ' and obtain the value zero as we did for $x < -1$. Then

$$T_F^{-1}\left\{\frac{2\sin\alpha}{\alpha}\right\} = \begin{cases} 0 & \text{if } |x| > 1 \\ 2\pi & \text{if } |x| < 1 \end{cases}$$

PROBLEM 7.2

Use the Fourier inversion formula to find $T_F^{-1}\{\alpha^{-1/2}\}$.

SOLUTION 7.2

This function $F(\alpha)$ is not square integrable. Nevertheless, the inversion formula can be applied to find a function $f(x)$, also not square integrable, whose Fourier transform equals $F(\alpha)$. We write

$$T_F^{-1}\{\alpha^{-1/2}\} = \int_{-\infty}^{\infty} \frac{e^{ix\alpha}}{\sqrt{\alpha}} d\alpha \tag{1}$$
$$= \lim_{R \to \infty} \int_{\Gamma_R} \frac{e^{ixz}}{\sqrt{z}} dz$$

If $x < 0$, then we may choose Γ_R to be the semicircular contour shown in Figure 7.1(a) and use Jordan's lemma to show that the integral over the curved part of the path tends to zero as R tends to infinity. Since the contour encloses no singularities of the integrand, the contour integral equals zero; i.e.,

$$T_F^{-1}\{\alpha^{-1/2}\}(x) = 0 \quad \text{for } x < 0. \tag{2}$$

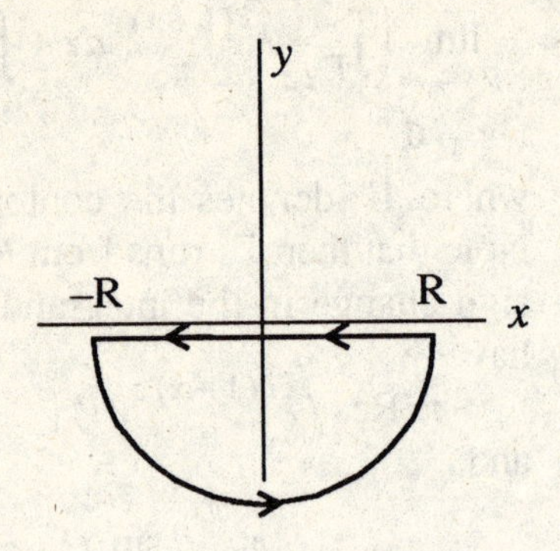

Figure 7.1(a)

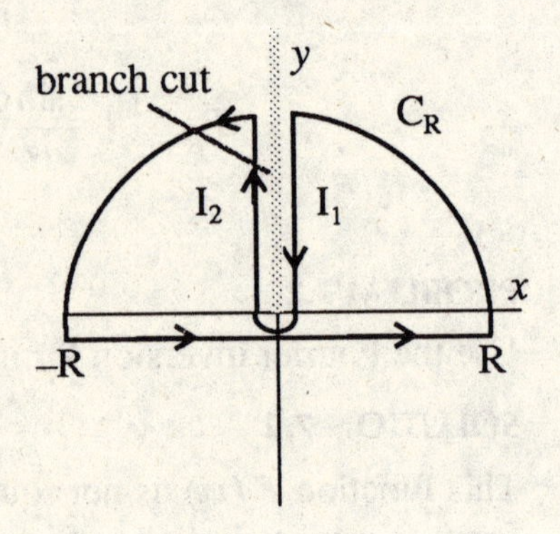

Figure 7.1(b)

If $x > 0$, we choose Γ_R to be the contour shown in Figure 7.1(b). Note that the branch cut for $z^{1/2}$ has been placed on the positive imaginary axis and the contour Γ_R encloses but does not touch this branch cut. Placement of the branch cut on the positive imaginary axis is equivalent to specifying $-3\pi/2 < \vartheta < \pi/2$. Then on the two segments, I_1 and I_2 of Γ_R which run parallel to the branch cut, we have

$$\text{on } I_1\text{: } z = se^{i\pi/2},\ 0 < s < \infty, \text{ hence } \sqrt{z} = s^{1/2}e^{i\pi/4}$$

$$\text{on } I_2\text{: } z = se^{-i3\pi/2},\ 0 < s < \infty, \text{ hence } \sqrt{z} = s^{1/2}e^{-i3\pi/4}.$$

The integrals over the large semicircle C_R and the small circle C_ε tend to zero as R tends to infinity and ε tends to zero. Thus, since the contour Γ_R encloses no singular points of the integrand,

$$0 = \lim_{\substack{R\to\infty \\ \varepsilon\to 0}} \int_\Gamma \frac{e^{ixz}}{\sqrt{z}}\,dz = \int_{-\infty}^{\infty} \frac{e^{ix\alpha}}{\sqrt{\alpha}}\,d\alpha + \int_{\infty}^{0} s^{-1/2}e^{-i\pi/4}e^{ix(is)}\,ids$$

$$+ \int_0^\infty s^{-1/2}e^{i3\pi/4}e^{ix(is)}\,ids$$

i.e.

$$\int_{-\infty}^{\infty} \frac{e^{ix\alpha}}{\sqrt{\alpha}}\,d\alpha = \int_0^\infty s^{-1/2}e^{-xs}\left(e^{-i\pi/4} - e^{i3\pi/4}\right)ids$$

$$= \left(e^{-i\pi/4} - e^{i3\pi/4}\right)i\int_0^\infty s^{-1/2}e^{-xs}\,ds$$

$$= 2e^{i\pi/4}\sqrt{\frac{\pi}{x}} \quad \text{for } x > 0 \tag{3}$$

Then (2) and (3) together give $T_F\{\alpha^{-1/2}\}$ for all x.

PROBLEM 7.3

Use the Fourier inversion formula to find $T_F^{\ -1}\{(\alpha^2+1)^{-1/2}\}$.

SOLUTION 7.3

The inversion formula asserts that

$$T_F^{\ -1}\{(\alpha^2+1)^{-1/2}\} = \int_{-\infty}^{\infty} \frac{e^{ix\alpha}}{\sqrt{\alpha^2+1}}\,d\alpha$$

The integrand

$$G(z) = \frac{e^{ixz}}{\sqrt{z^2+1}} = \frac{e^{ixz}}{\sqrt{z-i}\sqrt{z+i}}$$

has branch points at $z = i, -i$. If we choose branch cuts for the two double valued functions in the denominator as follows

$$-3\pi/2 < \text{Arg}\,(z-i) < \pi/2 \quad \text{and} \quad -3\pi/2 < \text{Arg}\,(z+i) < \pi/2 \tag{1}$$

then it is not difficult to check that the product $(z^2+1)^{1/2}$ is single valued everywhere except on the segment of the imaginary axis from $-i$ to i. See Figure 7.2. Then this segment is a branch cut for $G(z)$. At points $z = is$ lying just on the right side of the branch cut we have $\vartheta_1 = \text{Arg}\,(z-i) = -\pi/2$, $\vartheta_2 = \text{Arg}\,(z+i) = \pi/2$ and

$$\sqrt{z^2+1} = \left|(z^2+1)^{1/2}\right| e^{i\vartheta_1/2} e^{i\vartheta_2/2}$$

$$= \sqrt{1-s^2}\, e^{i(-\pi/2+\pi/2)/2} = \sqrt{1-s^2}$$

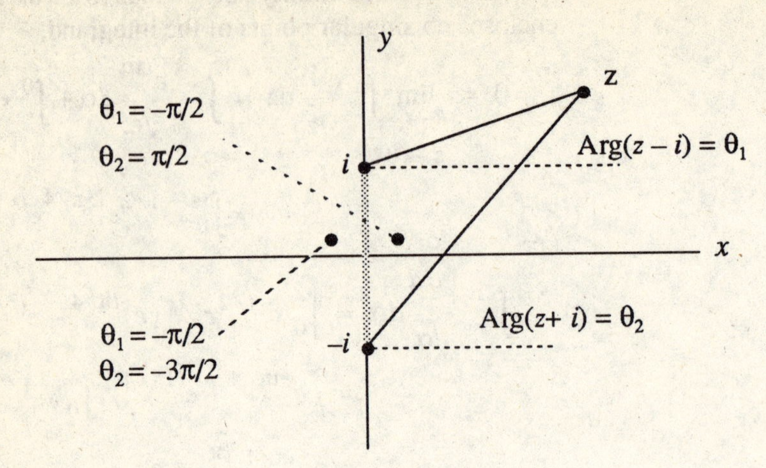

Figure 7.2

At points on the opposite side of the branch cut, at points $z = is$ lying on the left side of the cut, we have $\vartheta_1 = -3\pi/2$ and $\vartheta_2 = -\pi/2$. Therefore, at these points

$$\sqrt{z^2+1} = \left|(z^2+1)^{1/2}\right| e^{i\vartheta_1/2} e^{i\vartheta_2/2}$$

$$= \sqrt{1-s^2}\, e^{i(-3\pi/2-\pi/2)/2} = \sqrt{1-s^2}\, e^{-i\pi}$$

$$= -\sqrt{1-s^2}$$

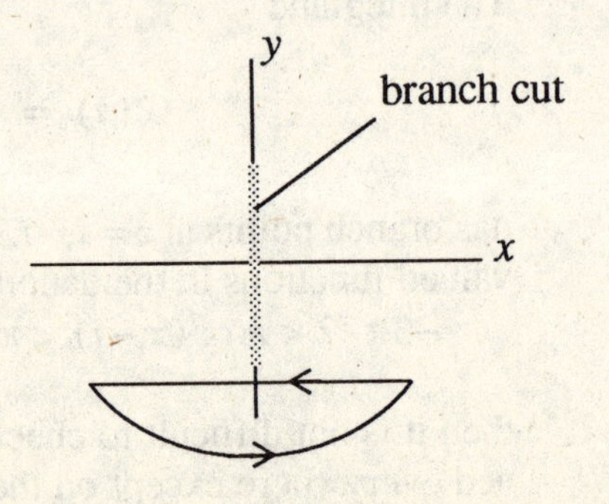

Figure 7.3(a)

If $x < 0$, we may choose Γ_R to be the contour shown in Figure 7.3(a) and use Jordan's lemma to show that the integral over the semicircular part of the contour tends to zero as R tends to infinity. Since Γ_R encloses no singular points of the integrand, (1) has the value zero when $x < 0$.

If $x > 0$ we choose Γ_R to be the contour shown in Figure 7.3(b).

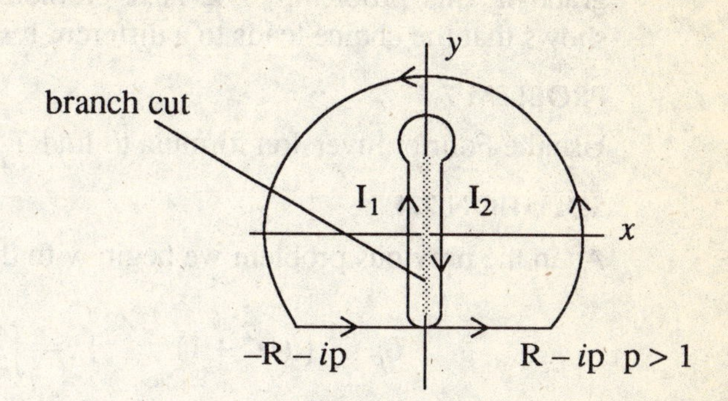

Figure 7.3(b)

Since the contour encloses no singular points of the integrand, the integral over this path equals zero. In addition, the integrals over the curved parts of the contour can be shown to tend to zero as R tends to infinity and ε tends to zero. This leaves

$$\int_{-\infty}^{\infty} ((q-ip)^2+1)^{-1/2} e^{ix(q-ip)}\,dq + \int_{-1}^{1} \frac{-1}{\sqrt{1-s^2}} e^{ix(is)}\,i\,ds$$

$$+ \int_{1}^{-1} \frac{1}{\sqrt{1-s^2}} e^{ix(is)}\,i\,ds = 0;$$

i.e., letting $\alpha = q - ip$,

$$\int_{-\infty}^{\infty} \frac{e^{ix\alpha}}{\sqrt{\alpha^2+1}}\,d\alpha = 2i\int_{-1}^{1} \frac{e^{-xs}}{\sqrt{1-s^2}}\,ds \tag{2}$$

If we let $s = \sin\varphi$, the integral is transformed to a well known integral representation for a modified Bessel function; i.e.,

$$\int_{-\infty}^{\infty} \frac{e^{ix\alpha}}{\sqrt{\alpha^2+1}}\,d\alpha = 2i\int_{-\pi/2}^{\pi/2} e^{-x\sin\varphi}\,d\varphi = 2\pi i I_0(x) \tag{3}$$

where $I_0(x)$ denotes the modified Bessel function of the first kind of order zero. Thus

$$T_F^{-1}\{(\alpha^2+1)^{-1/2}\} = \begin{cases} 0 & \text{if } x<0 \\ 2\pi i I_0(x) & \text{if } x>0 \end{cases}$$

Note that (1) is just one possible choice for the branch cut for the integrand in this problem. The next problem uses a different choice and shows that the choice leads to a different result for the inverse transform.

PROBLEM 7.4

Use the Fourier inversion formula to find $T_F^{-1}\{(\alpha^2+1)^{-1/2}\}$.

SOLUTION 7.4

As in the previous problem we begin with the inversion formula

$$T_F^{-1}\{(\alpha^2+1)^{-1/2}\} = \int_{-\infty}^{\infty} \frac{e^{ix\alpha}}{\sqrt{\alpha^2+1}}\,d\alpha$$

However, we choose a different branch cut for the multiple valued integrand

$$G(z) = \frac{e^{ixz}}{\sqrt{z^2+1}} = \frac{e^{ixz}}{\sqrt{z-i}\sqrt{z+i}},$$

in specifying

$$-3\pi/2 < \vartheta_1 = \operatorname{Arg}(z-i) < \pi/2$$

and (2)

$$-\pi/2 < \vartheta_2 = \operatorname{Arg}(z+i) < 3\pi/2$$

Then it is easy to verify that $G(z)$ is single valued at all points in the complex plane except points of the form $z = iy$, with $|y| > 1$, (see Figure 7.4).

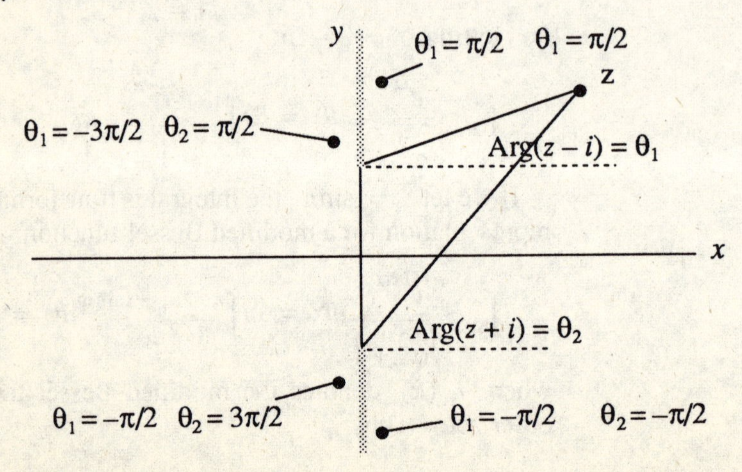

Figure 7.4

If $x < 0$, then to compute the inverse transform by means of a contour integral, we choose a contour Γ_R in the lower half plane as shown in Figure 7.5(a). This contour avoids that part of the branch cut in the lower half plane and encloses no singular points of the integrand $G(z)$. In addition, the integral over the large semicircle C_R tends to zero as R tends to infinity and the integral over C_ε, the small circle around the branch point at $z = -i$, tends to zero as ε tends to zero.

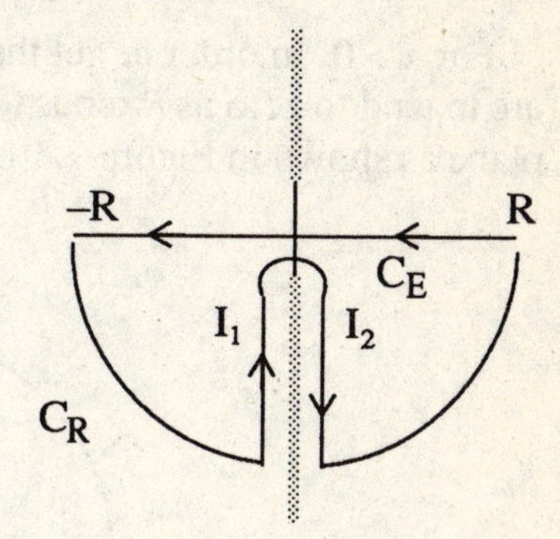

Figure 7.5(a)

That is,

$$\begin{aligned} 0 &= \int_{\Gamma_R} G(z)\,dz \\ &= \int_R^{-R} G(z)\,dz + \int_{C_R} G(z)\,dz + \int_{I_1} G(z)\,dz + \int_{C_\varepsilon} G(z)\,dz + \int_{I_2} G(z)\,dz \end{aligned}$$

and, letting R tend to infinity and ε tend to zero, we are left with

$$0 = \int_{\infty}^{-\infty} G(\alpha)\,d\alpha + 0 + \int_{-\infty}^{-1} G(z)\,dz + \int_{-1}^{-\infty} G(z)\,dz. \tag{3}$$

At points lying on the part I_1 of the contour we have $\vartheta_1 = \operatorname{Arg}(z-i) = -\pi/2$, $\vartheta_2 = \operatorname{Arg}(z+i) = 3\pi/2$ and $z = is$. Then

$$\begin{aligned} \sqrt{z^2+1} &= \left|(z^2+1)^{1/2}\right| e^{i\vartheta_1/2} e^{i\vartheta_2/2} \\ &= \sqrt{s^2-1}\,e^{i(-\pi/2+3\pi/2)/2} = \sqrt{s^2-1}\,e^{i\pi/2} \end{aligned}$$

Similarly, for points on I_2 we have $\vartheta_1 = -\pi/2$ and $\vartheta_2 = -\pi/2$, hence at these points

$$\begin{aligned} \sqrt{z^2+1} &= \left|(z^2+1)^{1/2}\right| e^{i\vartheta_1/2} e^{i\vartheta_2/2} \\ &= \sqrt{s^2-1}\,e^{i(-\pi/2-\pi/2)/2} = \sqrt{s^2-1}\,e^{-i\pi/2} \end{aligned}$$

Using these results in (3) leads to

$$\int_{-\infty}^{\infty}\frac{e^{ix\alpha}}{\sqrt{\alpha^2+1}}d\alpha = \int_{-\infty}^{-1}\frac{ie^{-xs}}{\sqrt{s^2-1}}ids+\int_{-1}^{-\infty}\frac{ie^{-xs}}{\sqrt{s^2-1}}ids$$

$$= -2\int_{-1}^{-\infty}\frac{e^{-xs}}{\sqrt{s^2-1}}ds = 2\int_{1}^{\infty}\frac{e^{xs}}{\sqrt{s^2-1}}ds \quad \text{for } x<0. \tag{4}$$

For $x>0$, in order to get the integral of $G(z)$ over the large semicular arc to tend to zero as R tends to infinity, we use a contour in the upper half plane as shown in Figure 7.5(b).

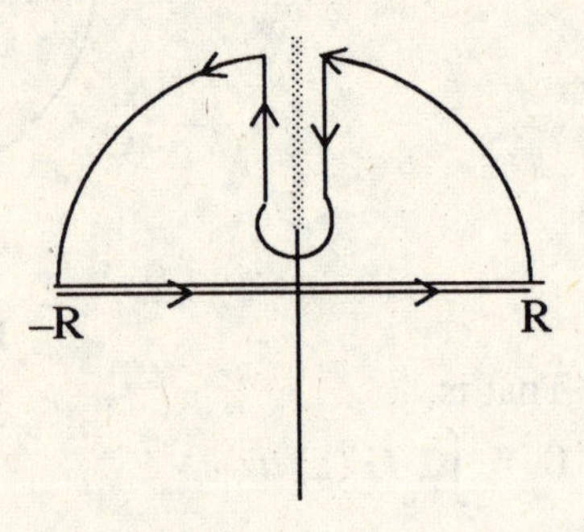

Figure 7.5(b)

Then in much the same way, we find

$$\int_{-\infty}^{\infty}\frac{e^{ix\alpha}}{\sqrt{\alpha^2+1}}d\alpha = 2\int_{1}^{\infty}\frac{e^{-xs}}{\sqrt{s^2-1}}ds \quad \text{for } x>0. \tag{5}$$

The results (4) and (5) together, imply

$$\int_{-\infty}^{\infty}\frac{e^{ix\alpha}}{\sqrt{\alpha^2+1}}d\alpha = 2\int_{1}^{\infty}\frac{e^{-|x|s}}{\sqrt{s^2-1}}ds$$

The change of variable, $s = \cosh\varphi$, reduces this integral to a well known integral representation for a modified Bessel function of the second kind; i.e.,

$$T_F^{-1}\left\{\frac{1}{\sqrt{\alpha^2+1}}\right\} = 2\int_0^{\infty}e^{-|x|\cosh\varphi}d\varphi = 2K_0(|x|)$$

where $K_0(x)$ denotes the modified Bessel function of the second kind of order zero.

PROBLEM 7.5

Find a square integrable function $u = u(x)$ such that u' and u'' are also square integrable and

$$xu'' + u'(x) - xu(x) = 0. \tag{1}$$

SOLUTION 7.5

If we let $U(\alpha) = T_F\{u(x)\}$, then $T_F\{u'(x)\} = i\alpha U(\alpha)$ by property 3 in Theorem 7.3. Using property 3 together with property 4 implies

$$T_F\{xu(x)\} = iU'(\alpha) \text{ and } T_F\{xu''(x)\} = id/d\alpha(-\alpha^2 U(\alpha)).$$

Then (1) becomes

$$id/d\alpha(-\alpha^2 U(\alpha)) + i\alpha U(\alpha) - iU'(\alpha) = 0$$

or

$$(1+\alpha^2)U'(\alpha) + \alpha U(\alpha) = 0. \tag{2}$$

Equation (2) separates and is easily solved to yield

$$U(\alpha) = \frac{C}{\sqrt{\alpha^2+1}} \tag{3}$$

for C an arbitrary constant of integration. Inverting the transform using the contour from Problem 7.3 gives the result

$$u_1(x) = C_1 I_0(x)$$

while inverting using the contour from Problem 7.4 leads to

$$u_2(x) = C_2 K_0(|x|).$$

Thus the general solution for (1) is a linear combination of $u_1(x)$ and $u_2(x)$. It should be noted, however, that the modified Bessel function of the second kind grows without bound as x tends to zero.

PROBLEM 7.6

Find a square integrable function $u = u(x, y)$ satisfying

$$\partial_{xx}u(x,y) + \partial_{yy}u(x,y) = 0 \text{ for } -\infty < x < \infty, y > 0 \tag{1}$$

$$u(x,0) = f(x), \quad u(x,y) \text{ bounded for all } x, y > 0. \tag{2}$$

Here $f(x)$ denotes a given square integrable function.

SOLUTION 7.6

If we let $U(\alpha, y)$ denote the Fourier transform in the x-variable of $u(x, y)$, then $T_F\{\partial_{xx}u\} = -\alpha^2 U(\alpha, y)$ and $T_F\{\partial_{yy}u\} = \partial_{yy}U(\alpha, y)$; i.e., since the transform is in the variable x, the derivative with respect to x interacts with the transform acoording to property 3 of Theorem 7.3 but the y-derivative does not interact with the transform. Then (1) and (2) become

$$-\alpha^2 U(\alpha, y) + \partial_{yy} U(\alpha, y) = 0, \quad U(\alpha, 0) = F(\alpha),$$
$$U(\alpha, y) \text{ bounded for } y > 0.$$

This simple ordinary differential equation in y has the general solution

$$U(\alpha, y) = C_1 e^{-y|\alpha|} + C_2 e^{y|\alpha|} \tag{3}$$

Clearly $C_2 = 0$ if $U(\alpha, y)$ is to remain bounded for all $y > 0$, and if $U(\alpha, 0) = F(\alpha)$ then $C_1 = F(\alpha)$. Then

$$U(\alpha, y) = F(\alpha) e^{-y|\alpha|} \tag{4}$$

Now Example 7.1(b) and property 8 of Theorem 7.3 imply that

$$u(x, y) = \frac{1}{2\pi}\left(\frac{2y}{x^2 + y^2}\right) * f(x) = \frac{1}{2\pi}\int_{-\infty}^{\infty} \frac{2yf(\xi)}{(x-\xi)^2 + y^2} d\xi \tag{5}$$

Then (5) is the solution of the boundary value problem (1), (2).

PROBLEM 7.7

Find a square integrable function $u = u(x, y)$ satisfying

$$\partial_{xx} u(x, y) + \partial_{yy} u(x, y) = 0 \text{ for } -\infty < x < \infty, 0 < y < a \tag{1}$$

$$u(x, 0) = f(x), \quad \partial_y u(x, a) = 0 \quad \text{for } -\infty < x < \infty. \tag{2}$$

Here $f(x)$ denotes a given square integrable function.

SOLUTION 7.7

Proceeding as in the previous problem, (1), (2) transforms to the following boundary value problem for an ordinary differential equation in the unknown function $U(\alpha, y)$

$$-\alpha^2 U(\alpha, y) + \partial_{yy} U(\alpha, y) = 0, U(\alpha, 0) = F(\alpha), \partial_y U(\alpha, a) = 0.$$

The general solution for this differential equation can be written in the form

$$U(\alpha, y) = C_1 \sinh \alpha y + C_2 \cosh \alpha (a - y) \qquad 0 < y < a.$$

Then $U(\alpha, 0) = C_2 \cosh \alpha a = F(\alpha)$ implies $C_2 = F(\alpha) / \cosh \alpha a$. Similarly, $\partial_y U(\alpha, a) = C_1 \cosh \alpha a = 0$ leads to $C_1 = 0$. Then

$$U(\alpha, y) = F(\alpha)\frac{\cosh\alpha(a-y)}{\cosh\alpha a} \tag{3}$$

Now

$$T_F^{-1}\{\frac{\cosh\alpha(a-y)}{\cosh\alpha a}\} = \int_{-\infty}^{\infty} e^{ix\alpha}\frac{\cosh\alpha(a-y)}{\cosh\alpha a}d\alpha$$

$$= \frac{1}{2}\int_{-\infty}^{\infty}\frac{1}{\cosh\alpha a}(e^{\alpha(ix+a-y)} + e^{\alpha(ix-a+y)})\,d\alpha$$

and therefore we consider an integral of the form

$$I_b = \int_\Gamma \frac{e^{bz}}{\cosh az}dz \quad \text{for } b = \text{complex constant} \tag{4}$$

Proceeding as we did in Problem 5.20 we can use the residue theorem to show

$$I_b = \frac{\pi}{a}\frac{1}{\cosh(\pi b/2a)}$$

hence

$$T_F^{-1}\{\frac{\cosh\alpha(a-y)}{\cosh\alpha a}\}$$

$$= \frac{\pi}{2a}\left(\frac{1}{\cos(\pi(ix+a-y)/2a)} + \frac{1}{\cos(\pi(ix-a+y)/2a)}\right)$$

$$= \frac{\pi}{2a}\left(\frac{1}{\cosh(\pi x/2a)\cos(\pi(a-y)/2a) - \sinh(\pi x/2a)\sin(\pi(a-y)/2a)}\right.$$

$$\left. + \frac{1}{\cosh(\pi x/2a)\cos(\pi(a-y)/2a) + \sinh(\pi x/2a)\sin(\pi(a-y)/2a)}\right)$$

$$= \frac{\pi}{a}\frac{\cos(\pi(a-y)/2a)\cosh(\pi x/2a)}{(\cosh(\pi x/2a)\cos(\pi(a-y)/2a))^2 + (\sinh(\pi x/2a)\sin(\pi(a-y)/2a)}$$

But

$$(\cos A\cosh B)^2 + (\sin A\sinh B)^2 = \cos^2 A + \sinh^2 B$$

$$= \frac{1}{2}(\cos 2A + \sinh 2B)$$

and thus

$$T_F^{-1}\{\frac{\cosh\alpha(a-y)}{\cosh\alpha a}\} = \frac{2\pi}{a}\frac{\cos(\pi(a-y)/2a)\cosh(\pi x/2a)}{\cos(\pi(a-y)/a) + \sinh(\pi x/a)}$$

Finally, then property 8 of Theorem 7.3 implies that the solution of the boundary value problem is given by

$$u(x, y) = \frac{1}{a}\int_{-\infty}^{\infty} \frac{\cos(\pi(a-y)/2a)\cosh(\pi(x-\xi)/2a)}{\cos(\pi(a-y)/a) + \sinh(\pi(x-\xi)/2a)} f(\xi)\, d\xi$$

The Laplace Transform

PROBLEM 7.8

Use the inversion formula (7.6) for the Laplace transform to find the function $f(t)$ whose Laplace transform is:

(a) $F(s) = 1/(s-a)$ for positive constant a.

(b) $F(s) = 1/(s^2 + a^2)$

(c) $F(s) = 1/(s^2 + a^2)^2$

SOLUTION 7.8

The function $F(z) = 1/(z-a)$ has a single simple pole at $z = a$. Let Γ denote the simple closed contour shown in Figure 7.6 consisting of the curved arc C_R and the straight line from $c - ip$ to $c + ip$ with $c > a$.

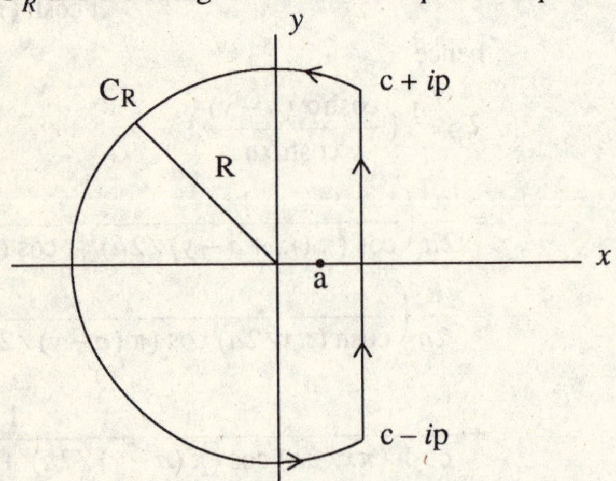

Figure 7.6

Then the pole lies inside Γ and

$$\int_{\Gamma} e^{zt}(z-a)\, dz = \int_{C_R} e^{zt}(z-a)^{-1} dz + \int_{c-ip}^{c+ip} e^{zt}(z-a)^{-1} dz$$

$$= 2\pi i \text{es}\, [e^{zt}]\Big|_{z=a} = 2\pi i e^{at}.$$

The equation (1) holds for all such contours Γ and as R tends to infinity, we can show in the usual way that the integral of $e^{zt}F(z)$ over C_R tends to zero. Then it follows that

$$f(t) = \frac{1}{2\pi i} \lim_{p \to \infty} \int_{c-ip}^{c+ip} e^{zt}(z-a)^{-1} dz = e^{at} \tag{2}$$

The second function

$$F(z) = 1/(z^2+a^2) = \frac{1}{(z-ia)(z+ia)}$$

has two simple poles located at $z = ai, -ai$. Then, using the same contour, Γ, we get

$$\int_\Gamma e^{zt}F(z)\,dz = \int_{C_R} e^{zt}F(z)\,dz + \int_{c-ip}^{c+ip} e^{zt}F(z)\,dz$$

$$= 2\pi i\left(\text{Res}\,(e^{zt}F(z))\Big|_{z=-ia} + \text{Res}\,(e^{zt}F(z))\Big|_{z=ia}\right)$$

$$= 2\pi i\left((e^{zt}/z-ia)\Big|_{z=-ia} + (e^{zt}/z+ia)\Big|_{z=ia}\right)$$

$$= 2\pi i\left(\frac{e^{-iat}}{-2ai} + \frac{e^{iat}}{2ai}\right) = 2\pi i\left(\frac{\sin at}{a}\right).$$

It follows from this computation that

$$f(t) = T_L^{-1}\left\{\frac{1}{s^2+a^2}\right\} = \frac{\sin at}{a}$$

Finally, the function $F(z) = 1/(z^2+a^2)^2$ has poles of order two at $z = -ia, ia$ and thus, proceeding as in the previous case, we find

$$\int_\Gamma e^{zt}F(z)\,dz$$

$$= 2\pi i\left(\text{Res}\left\lfloor e^{zt}/(z^2+a^2)^2\right\rfloor\Big|_{z=-ia} + \text{Res}\left\lfloor e^{zt}/(z^2+a^2)^2\right\rfloor\Big|_{z=ia}\right)$$

$$= 2\pi i\left(e^{-iat}\frac{-2tai-2}{(-2ai)^3} + e^{iat}\frac{2tai-2}{(-2ai)^3}\right)$$

Then

$$\lim_{R\to\infty}\left(\int_{C_R} e^{zt}F(z)\,dz + \int_{c-ip}^{c+ip} e^{zt}F(z)\,dz\right)/2\pi i = f(t)$$

$$= T_L^{-1}\left\{\frac{1}{(s^2+a^2)^2}\right\} = \frac{1}{2a^3}(\sin at - at\cos at)$$

PROBLEM 7.9

Use the inversion formula (7.6) for the Laplace transform to find the function $f(t)$ whose Laplace transform is

$$F(s) = \frac{e^{-b\sqrt{s}}}{\sqrt{s}} \quad \text{for } b > 0. \tag{1}$$

SOLUTION 7.9

The function $F(z) = z^{-1/2}e^{-b\sqrt{z}}$ is double valued with a branch point at $z = 0$. If we place the branch cut along the negative real axis, corresponding to the range $-\pi < \vartheta < \pi$ for ϑ, then $e^{zt}F(z)$ is analytic everywhere except at points on the branch cut. The contour Γ for the inversion integral can be modified as shown in Figure 7.7 to consist of the vertical line from $c - ip$ to $c + ip$, with $c > 0$, the two circular arcs C_R and C_ε, and the two straight segments I_1 and I_2. The lines I_1 and I_2 lie, respectively, just above and just below the branch cut and thus for all positive R and ε, the integral of $e^{zt}F(z)$ over Γ is equal to zero.

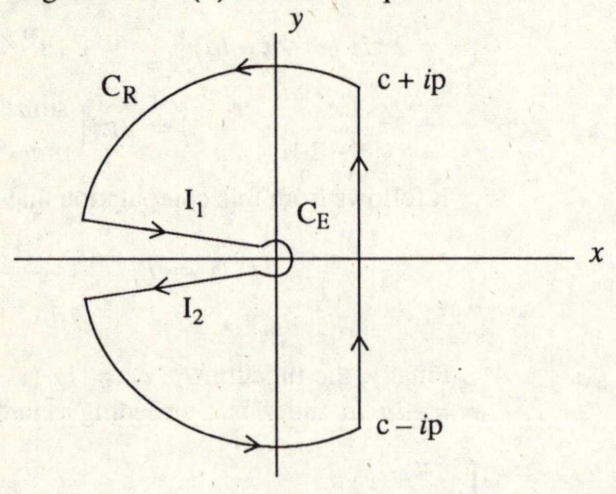

Figure 7.7

We have

$$\text{on } I_1\text{: } z = re^{i\pi} = -r \text{ and } \sqrt{z} = ir^{1/2} \text{ for } \varepsilon < r < R$$

$$\text{on } I_2\text{: } z = re^{-i\pi} = -r \text{ and } \sqrt{z} = -ir^{1/2} \text{ for } \varepsilon < r < R$$

and thus

$$0 = \int_\Gamma e^{zt}F(z)\,dz = \int_{c-ip}^{c+ip} e^{zt}F(z)\,dz + \int_{C_R} e^{zt}F(z)\,dz + \int_{C_\varepsilon} e^{zt}F(z)\,dz$$
$$- \int_\varepsilon^R e^{-rt}\left(F(re^{-i\pi}) - F(re^{i\pi})\right)dr.$$

For $F(z) = z^{-1/2}e^{-b\sqrt{z}}$, $b > 0$, we can show that the integrals over C_R and C_ε tend to zero as R tends to infinity and ε tends to zero. Since the integral over Γ is zero for all positive R and ε, this leaves

$$\lim_{p\to\infty}\int_{c-ip}^{c+ip} e^{zt}F(z)\,dz = \int_0^\infty e^{-rt}\left(F(re^{-i\pi}) - F(re^{i\pi})\right)dr \qquad (2)$$

Now

$$\int_0^\infty e^{-rt}(F(re^{-i\pi}) - F(re^{i\pi}))\,dr = \int_0^\infty e^{-rt}\left(\frac{e^{ib\sqrt{r}}}{i\sqrt{r}} - \frac{e^{-ib\sqrt{r}}}{i\sqrt{r}}\right)dr$$

$$= 2i\int_0^\infty e^{-rt}\frac{\cos b\sqrt{r}}{\sqrt{r}}\,dr$$

and so it follows from (2) that

$$f(t) = \frac{1}{2\pi i}\lim_{p\to\infty}\int_{c-ip}^{c+ip} e^{zt}F(z)\,dz = \frac{1}{\pi}\int_0^\infty e^{-rt}\frac{\cos b\sqrt{r}}{\sqrt{r}}\,dr \qquad (3)$$

Letting $\lambda = \sqrt{r}$ in (3) produces the result

$$f(t) = \frac{2}{\pi}\int_0^\infty e^{-t\lambda^2}\cos b\lambda\,d\lambda$$

$$= \frac{1}{\pi}\int_{-\infty}^\infty e^{-t\lambda^2}e^{ib\lambda}\,d\lambda = \frac{1}{\pi}e^{-b^2/4t}\int_{-\infty}^\infty e^{-t(\lambda - ib/2t)^2}\,d\lambda$$

$$= \frac{1}{\sqrt{\pi t}}e^{-b^2/4t}.$$

Thus

$$T_F^{-1}\left\{\frac{e^{-b\sqrt{s}}}{\sqrt{s}}\right\} = \frac{1}{\sqrt{\pi t}}e^{-b^2/4t} \quad \text{for } b > 0.$$

PROBLEM 7.10

Use the inversion formula (7.6) for the Laplace transform to find the function $f(t)$ whose Laplace transform is

$$F(s) = \frac{e^{-b\sqrt{s}}}{s} \quad \text{for } b > 0. \qquad (1)$$

SOLUTION 7.10

To find the inverse transform for $F(s)$ we use the same branch cut and contour as in the previous problem. The denominator contains a higher power of s than the function in the previous problem and thus the integral of $e^{zt}F(z)$ over C_R tends to even more rapidly to zero as R tends to infinity. However the change in the denominator also produces the result that

$$\int_{C_\varepsilon} e^{zt}F(z)\,dz = -2\pi i \quad \text{for all } \varepsilon > 0 \qquad (2)$$

Also

$$\int_{I_1} e^{zt}F(z)\,dz + \int_{I_2} e^{zt}F(z)\,dz = \int_\varepsilon^R e^{-rt}\left(\frac{e^{ib\sqrt{r}}}{-r} - \frac{e^{-ib\sqrt{r}}}{-r}\right)dr$$

$$= -2i\int_{\varepsilon}^{R} e^{-rt}\frac{\sin b\sqrt{r}}{r}dr$$

Then it follows as before that

$$f(t) = 1 - \frac{1}{\pi}\int_{0}^{\infty} e^{-rt}\frac{\sin b\sqrt{r}}{r}dr \tag{3}$$

The change of variable $\lambda = \sqrt{r}$ reduces (3) to

$$f(t) = 1 - \frac{1}{\pi}\int_{0}^{\infty} e^{-t\lambda^2}\frac{\sin b\lambda}{\lambda}d\lambda \tag{4}$$

Then

$$f(t) = 1 - \frac{1}{\pi}\int_{0}^{\infty} e^{-t\lambda^2}\int_{0}^{b}\cos p\lambda\, dp\, d\lambda$$

$$= 1 - \frac{1}{\pi}\int_{0}^{b}\int_{0}^{\infty} e^{-t\lambda^2}\cos p\lambda\, dp\, d\lambda$$

$$= 1 - \frac{1}{\pi}\int_{0}^{b}\sqrt{\frac{\pi}{4t}}e^{-p^2/4t}dp$$

where we have used the result of the previous problem in evaluating the integral with respect to λ. Now, letting $q = p/\sqrt{4t}$ brings this integral into the form

$$f(t) = 1 - \pi^{-1/2}\int_{0}^{b/\sqrt{4t}} e^{-q^2}dq = \text{erfc}\,(\frac{b}{\sqrt{4t}})$$

Here erfc (x) denotes the *complementary error function*; i.e.,

$$T_F^{-1}\{\frac{e^{-b\sqrt{s}}}{s}\} = \text{erfc}\,(\frac{b}{\sqrt{4t}}) \quad \text{for } b > 0.$$

PROBLEM 7.11

Use the inversion formula (7.6) for the Laplace transform to find the function $f(t)$ whose Laplace transform is

$$F(s) = \frac{\sinh a\sqrt{s}}{s\sinh\sqrt{s}} \quad \text{for } a > 0 \tag{1}$$

SOLUTION 7.11

By writing

$$F(z) = \frac{\sinh a\sqrt{z}}{z\sinh\sqrt{z}} = \frac{a\sqrt{z} + (a\sqrt{z})^3/3! + (a\sqrt{z})^5/5! + \dots}{z(\sqrt{z} + (\sqrt{z})^3/3! + (\sqrt{z})^5/5! + \dots)}$$

$$= \frac{a + (az)^3/3! + (az)^5/5! + \ldots}{z(1 + z^3/3! + z^5/5! + \ldots)}$$

and noting that the power series in the numerator and denominator converge for all z, it becomes clear that $F(z)$ is analytic for all z where $\sinh\sqrt{z}$ is different from zero; i.e., $F(z)$ is analytic except at $z_n = -(n\pi)^2$ for all integer values of n. Note that $F(z)$ has no branch cut so long as the same branch of $\sqrt{z}$ is chosen for the numerator and denominator of $F(z)$. Each of the singularities z_n is a simple pole for $F(z)$ and all of the z_n lie in the half-plane, $\operatorname{Re} z \le 0$.

We can show that $F(z)$ satisfies the hypotheses of Theorem 7.5 and thus it follows that

$$f(t) = T_L\{F(s)\} = \frac{1}{2\pi i} \lim_{p \to \infty} \int_{c-ip}^{c+ip} e^{zt} F(z)\,dz$$

$$= \sum_{n=0}^{\infty} \operatorname{Res}\,[e^{zt}F(z)]\Big|_{z = z_n}$$

Now

$$\operatorname{Res}\,[e^{zt}F(z)]\Big|_{z=0} = \lim_{z \to 0} z e^{zt} F(z) = a$$

and

$$\operatorname{Res}\,[e^{zt}F(z)]\Big|_{z = -(n\pi)^2} = e^{-(n\pi)^2 t} \frac{p(z_n)}{q'(z_n)}$$

where

$$p(z) = z^{-1}\sinh a\sqrt{z} \quad \text{and} \quad q(z) = \sinh\sqrt{z}$$

Then

$$\frac{p(-n^2\pi^2)}{q'(-n^2\pi^2)} = \frac{i\sin n\pi a}{-n^2\pi^2} \frac{2n\pi i}{\cos n\pi} = (-i)^n \frac{\sin n\pi a}{n\pi}$$

and finally,

$$f(t) = a + \sum_{n=1}^{\infty} (-i)^n \frac{\sin n\pi a}{n\pi} e^{-n^2\pi^2 t} = T_L^{-1}\{\frac{\sinh a\sqrt{s}}{s\sinh\sqrt{s}}\} \quad \text{for } a > 0$$

PROBLEM 7.12

Use the inversion formula (7.6) for the Laplace transform to find the function $f(t)$ whose Laplace transform is

$$F(s) = \frac{e^{-|x|\sqrt{s^2 - b^2}}}{2\sqrt{s^2 - b^2}} \quad \text{for } b > 0 \tag{1}$$

SOLUTION 7.12

Here we take the branch cut for the multiple valued function $\sqrt{z^2 - b^2}$ to be the segment of the real axis joining $z = -b$ to $z = b$. Then $F(z)$ given by (1) is analytic everywhere except on the branch cut. If we take Γ_1 to be the contour shown in Figure 7.8(a), then the integral of $e^{zt}F(z)$ over Γ_1 is zero since this contour encloses no singular points of the integrand. For $t < b|x|$, we can show that as R tends to infinity, the integral of $e^{zt}F(z)$ over the portion C_R of the contour tends to zero.

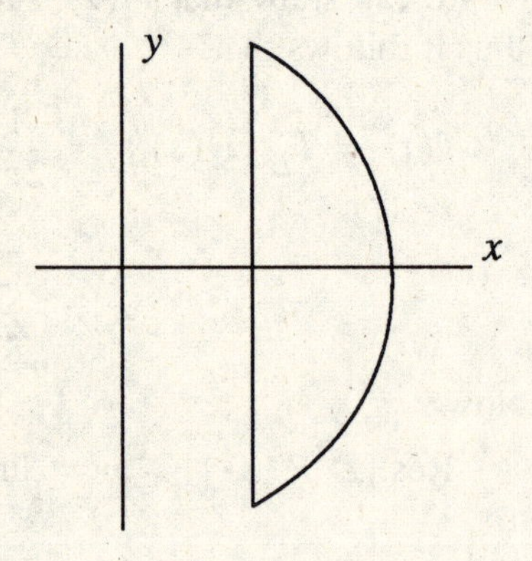

Figure 7.8(a)

This leaves

$$\int_{c - i\infty}^{c + i\infty} e^{zt} F(z)\, dz = 0$$

and then it follows from (7.6) that $f(t) = 0$ for $t < b|x|$.

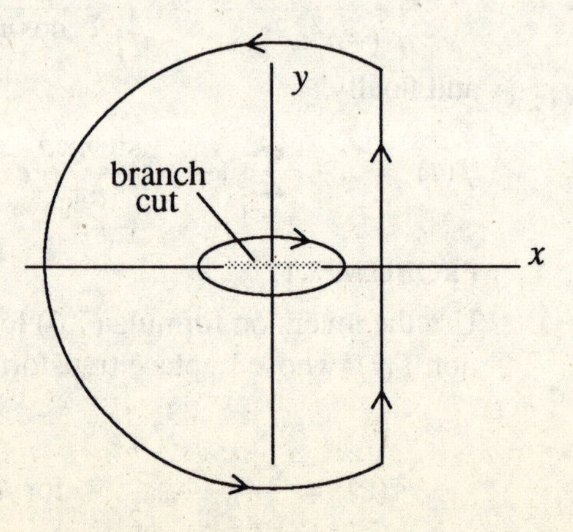

Figure 7.8(b)

For $t > b|x|$, we must use the contour Γ_2 shown in Figure 7.8(b); i.e., Γ_2 consists of the vertical line from $c - ip$ to $c + ip$, with $c > b$, the circular arc C_R, and the simple closed curve C_ε. Since the simple closed curve C_ε. Since the simple closed curve C_ε goes around the branch cut in the clockwise sense, the contour Γ_2 encloses no singular points of $e^{zt}F(z)$, hence the integral over Γ_2 is also zero. Furthermore, we can show that for $t > b|x|$, the integral of $e^{zt}F(z)$ over the circular arc, C_R, tends to zero as R tends to infinity. Then as C_ε shrinks down to just enclose the branch cut, we are left with

$$\int_{c-i\infty}^{c+i\infty} e^{zt}F(z)\,dz + \int_{b}^{-b} e^{st}F(-i\sqrt{b^2-s^2})\,ds + \int_{-b}^{b} e^{st}F(i\sqrt{b^2-s^2})\,ds = 0$$

Here we have used the fact that (see also Problem 7.3),

$$\sqrt{z^2-b^2} = i(b^2-s^2)^{1/2} \quad -b < s < b \text{ on the upper part of the branch cut}$$

$$\sqrt{z^2-b^2} = -i(b^2-s^2)^{1/2} \quad -b < s < b \text{ on the lower part of the branch cut}$$

It follows that

$$\int_{c-i\infty}^{c+i\infty} e^{zt}F(z)\,dz = \int_{-b}^{b} e^{st}\left(F(-i\sqrt{b^2-s^2}) - F(i\sqrt{b^2-s^2})\right)ds$$

$$= \int_{-b}^{b} e^{st}\left(\frac{\exp(i|x|\sqrt{b^2-s^2})}{-2i\sqrt{b^2-s^2}} - \frac{\exp(-i|x|\sqrt{b^2-s^2})}{2i\sqrt{b^2-s^2}}\right)ds$$

$$= i\int_{-b}^{b} e^{st}\frac{\cos(|x|\sqrt{b^2-s^2})}{\sqrt{b^2-s^2}}\,ds$$

Then (7.6) implies

$$f(t) = T_L^{-1}\{F(s)\} = \frac{1}{2\pi}\int_{-b}^{b} e^{st}\frac{\cos(|x|\sqrt{b^2-s^2})}{\sqrt{b^2-s^2}}\,ds$$

$$= \frac{1}{2\pi}\int_{-\pi/2}^{\pi/2} e^{bt\sin\varphi}\cos(b|x|\cos\varphi)\,d\varphi$$

where we used the change of variable, $s = b\sin\varphi$. Finally, we make use of an integral identity for the modified Bessel function of the first kind to write

$$f(t) = T_L^{-1}\{F(s)\} = \frac{1}{2b}I_0(\sqrt{t^2-b^2x^2})H(t-b|x|)$$

where $H(t)$ denotes the Heaviside step function

$$H(t) = \begin{cases} 0 & \text{if } t < 0 \\ 1 & \text{if } t > 0 \end{cases}$$

PROBLEM 7.13

For real constants a, A and given smooth function $f(t)$, use the Laplace transform to solve the following differential equations for $y(t)$:

(a) $y'(t) + ay(t) = f(t), \quad y(0) = 0$

(b) $y''(t) + a^2 y(t) = f(t), \quad y(0) = 0, y'(0) = 0$

(c) $y''(t) + a^2 y(t) = A \sin at$

SOLUTION 7.13

(a) If we let $Y(s)$ and $F(s)$ denote the Laplace transforms of $y(t), f(t)$ then it follows from property 3 in Theorem 7.6 that

$$sY(s) + aY(s) = (s+a)Y(s) = F(s);$$

i.e.,

$$Y(s) = \frac{1}{s+a} F(s).$$

Then the result of Problem 7.8(a), combined with property 7 of Theorem 7.6 leads to

$$y(t) = \int_0^t e^{-a(t-\tau)} f(\tau)\, d\tau.$$

(b) In the same way we find for equation (b)

$$s^2 Y(s) + a^2 Y(s) = (s^2 + a^2) Y(s) = F(s);$$

i.e.,

$$Y(s) = \frac{1}{s^2 + a^2} F(s).$$

Then

$$y(t) = \frac{1}{a} \int_0^t \sin a(t-\tau) f(\tau)\, d\tau.$$

(c) In equation (c) we have $f(t) = A \sin at$ and thus

$$F(s) = \frac{Aa}{s^2 + a^2} \quad \text{and} \quad Y(s) = \frac{Aa}{(s^2 + a^2)^2}$$

Then the result of Problem 7.8(c) leads to

$$y(t) = \frac{A}{2a^3} (\sin at - at \cos at)$$

PROBLEM 7.14

Use the Laplace transform to find a bounded function $u(x, t)$ such that

$$\partial_t u(x, t) = \partial_{xx} u(x, t) \quad \text{for } x > 0 \text{ and } t > 0,$$

$$u(x, 0) = 0 \quad \text{and} \quad u(0, t) = 1.$$

SOLUTION 7.14

Let $U(x, s)$ denote the Laplace transform of $u(x, t)$ with respect to t. Then

$$T_L\{\partial_t u(x, t)\} = sU(x, s) - 0 \text{ and } T_L\{\partial_{xx} u(x, t)\} = \partial_{xx} U(x, s)$$

and

$$sU(x, s) = \partial_{xx} U(x, s), \quad U(0, s) = 1/s.$$

Then we get

$$U(x, s) = C_1 e^{-x\sqrt{s}} + C_2 e^{x\sqrt{s}} \quad \text{for } x > 0.$$

To obtain the bounded solution, we choose $C_2 = 0$, and then the condition at $x = 0$ requires that $C_1 = 1/s$. Hence

$$U(x, s) = \frac{1}{s} e^{-x\sqrt{s}} \quad \text{for } x > 0$$

Then the result of Problem 7.10 implies that $u(x, t) = \text{erfc}\,(x/\sqrt{4t})$.

PROBLEM 7.15

Use the Laplace transform to find a bounded function $u(x, t)$ such that

$$\partial_t u(x, t) = \partial_{xx} u(x, t) \quad \text{for } x > 0 \text{ and } t > 0,$$

$$u(x, 0) = 0 \quad \text{and} \quad -\partial_x u(0, t) = g(t).$$

SOLUTION 7.15

Let $G(s)$ denote the Laplace transform of the given function $g(t)$. Then, proceeding as in the previous problem, we arrive at

$$U(x, s) = C_1 e^{-x\sqrt{s}}$$

The boundary condition at $x = 0$ requires that

$$-\partial_x U(0, s) = C_1 \sqrt{s} = G(s)$$

and thus

$$U(x, s) = \frac{1}{\sqrt{s}} e^{-x\sqrt{s}} G(s).$$

Then we use the result of Problem 7.9 with property 7 of Theorem 7.6 to conclude that

$$u(x,t) = \int_0^t \frac{1}{\sqrt{\pi(t-\tau)}} e^{-x^2/4(t-\tau)} g(\tau)\,d\tau$$

PROBLEM 7.16

Use the Laplace transform to find a function $u(x, t)$ such that

$$\partial_t u(x,t) = \partial_{xx} u(x,t) \quad \text{for } 0<x<1 \text{ and } t>0,$$

$$u(x,0) = 0 \text{ and } u(0,t) = 0 \text{ and } u(1,t) = A.$$

SOLUTION 7.16

In this problem, the differential equation transforms to

$$sU(x,s) = \partial_{xx} U(x,s) \qquad \text{for } 0<x<1.$$

The general solution to this equation may be written in the form

$$U(x,s) = C_1 \sin x\sqrt{s} + C_2 \sin(1-x)\sqrt{s}$$

and then the boundary conditions $U(0,s) = 0$ and $U(1,s) = A/s$ lead at once to

$$U(x,s) = \frac{A}{s}\frac{\sin x\sqrt{s}}{\sinh\sqrt{s}}.$$

Then the result of Problem 7.11 may be used to find the inverse transform

$$u(x,t) = Ax + A\sum_{n=1}^{\infty} \frac{(-1)^n}{n\pi} \sin n\pi x e^{-n^2\pi^2 t}$$

PROBLEM 7.17

Use the Laplace transform to find a bounded function $u(x, t)$ such that

$$\partial_{tt} u(x,t) = \partial_{xx} u(x,t) + b^2 u(x,t) + f(x,t) \quad \text{for all } x \text{ and } t>0,$$

$$u(x,0) = 0 \text{ and } \partial_t u(x,0) = 0.$$

SOLUTION 7.17

Let $U_1(x, s)$ and $F_1(x, s)$ denote the Laplace transform in t for the functions $u(x, t)$ and $f(x, t)$. Then the differential equation transforms to

$$s^2 U_1(x,s) - 0 = \partial_{xx} U_1(x,s) + b^2 U_1(x,s) + F_1(x,s)\,, \text{ for all } x.$$

Now let $U_2(\alpha, s)$ and $F_2(\alpha, s)$ denote the Fourier transform in x for the functions $U_1(x, s)$ and $F_1(x, s)$. Then

$$(s^2 + \alpha^2 - b^2)\, U_2(\alpha, s) = F_2(\alpha, s)$$

and

$$U_2(\alpha, s) = \frac{1}{s^2 - b^2 + \alpha^2} F_2(\alpha, s)\,.$$

Now the result of Example 7.2 shows that

$$T_F^{-1}\left\{\frac{1}{s^2 - b^2 + \alpha^2}\right\} = \frac{\pi}{\sqrt{s^2 - b^2}} e^{-|x|\sqrt{s^2 - b^2}}$$

and then the property 8 of Theorem 7.3 can be used to obtain

$$U_1(x, s) = \int_{-\infty}^{\infty} \frac{e^{-|x-\xi|\sqrt{s^2 - b^2}}}{2\sqrt{s^2 - b^2}} F_1(\xi, s)\, d\xi$$

Finally, we may use the result of Problem 7.12 together with property 7 of Theorem 7.6 to invert the Laplace transform and find

$$u(x,t) = \int_{-\infty}^{\infty} \int_{-\infty}^{t - b(x-\xi)} \frac{1}{2b} I_0\left(\sqrt{(t-\tau)^2 - b^2 (x-\xi)^2}\right) f(\xi, \tau)\, d\tau d\xi$$

PROBLEM 7.18

Determine the stability of the systems whose transfer functions are:

(a) $G_1(s) = 1/(s^4 + 2s^3 + s^2 + 3s + 2)$

(b) $G_2(s) = 1/(s^5 + s^4 + 7s^3 + 50s^2 + 1296s + 49)$

SOLUTION 7.18

We can use the argument principle to determine the number of zeros of $P_1(s) = s^4 + 2s^3 + s^2 + 3s + 2$ in the right half plane. We will compute the change in the argument of $P(z)$ as z traces out the path consisting of the semicircle $C_R = \{Re^{i\vartheta} : -\pi/2 < \vartheta < \pi/2\}$, followed by the straight line L from iR to $-iR$ along the imaginary axis. The result of Problem 5.22 implies that for R sufficiently large, all of the zeros of $P_1(s)$ in the right half plane will be contained inside this contour. Then by the argument principle it follows that the number of these zeros is equal to $\Delta \arg P_1(s)/2\pi$. If we write $P(iy) = y^4 - y^2 + 2 + iy(3 - 2y^2)$ then since $y^4 - y^2 + 2$ and $y(3 - 2y^2)$ have no common zeros, we see that $P_1(z)$ increases from -2π at $\vartheta = -\pi/2$ to 2π at $\vartheta = \pi/2$; i.e.,

$\Delta \arg P_1(s)(C_R) = 4(\pi/2) - 4(-\pi/2) = 4\pi$. Also

$$\tan(\arg P_1(iy)) = \frac{y(3-2y)^2}{y^4-y^2+2}$$

and thus as y tends to infinity, $\tan(\arg P_1(iy))$ tends to zero and $\varphi = \arg P_1(iy)$ tends to 2π. It is easy to check that $y^4 - y^2 + 2$ has no real zeros therefore as y decreases to the value $\sqrt{3/2}$, $\tan(\arg P_1(iy))$ first becomes negative then rises to the value zero. As y decreases from $\sqrt{3/2}$ to zero, $\tan\varphi$ goes positive and then returns to zero. Similarly, as y moves further down L from zero to $-\sqrt{3/2}$, $\tan\varphi$ goes negative and then returns to zero. Finally, as y continues down the imaginary axis, $\tan\varphi$ becomes positive but then decreases to zero as y tends to negative infinity. Thus $\Delta\varphi(L) = 2\pi - 2\pi = 0$ and the total change in $\arg P_1$ over the closed contour is 4π. It follows from the argument principle that P_1 has two roots in the right half plane and the system whose transfer function is $G_1(s) = 1/P_1(s)$ is not stable.

Applying the same reasoning to $P_2(s) = s^5 + s^4 + 97s^3 + 50s^2 + 1296s + 49$, we find first that $\Delta\varphi = \Delta(\arg P_2(Re^{i\vartheta})) = 5\pi$ as ϑ goes from $-\pi/2$ to $\pi/2$. Now $P_2(iy) = y^4 - 50y^2 + 49 + iy(y^4 - 97y^2 + 1296)$ and it is clear P_2 has no zeros on the imaginary axis. Furthermore,

$$\tan(\arg P_2(iy)) = \frac{y(y^2-16)(y^2-81)}{(y^2-1)(y^2-49)}$$

and as y moves down L, $\varphi = \arg(P_2(iy))$ varies as follows:

as $y \to \infty$ then $\tan\varphi \to \infty$ and $\varphi \to 5\pi/2$

as y decreases to 9 then $\tan\varphi$ decreases to 0 and $\varphi = 2\pi$ at $y = 9$

for $7 < y < 9$, $-\infty < \tan\varphi < 0$ and φ tends to $3\pi/2$ as y decreases to 7

for $4 < y < 7$, $0 < \tan\varphi < \infty$ as we move to a different branch of $\tan\varphi$

as y decreases to 4, $\tan\varphi$ decreases to zero and $\varphi = \pi$ at $y = 4$

as y decreases from 4 to 1, $\tan\varphi$ tends to $-\infty$ and φ decreases to $\pi/2$

as y decreases from 1 to 0, $\tan\varphi$ decreases from ∞ to 0 along a new branch of $\tan\varphi$. Then $\varphi = 0$ at $y = 0$.

Similarly, as $z = iy$ traces out the negative half of L, φ goes from 0 to $-5\pi/2$ and we have then $\Delta\varphi(L) = -5\pi/2 - (5\pi/2) = -5\pi$. Therefore the total change in $\varphi = \arg P_2$ over the closed contour it $\Delta\varphi = 5\pi - 5\pi = 0$; i.e., there are no zeros for $P_2(s)$ in the right half plane and the system whose transfer function is $G_2(s)$ is stable.

SUMMARY

The Fourier transform of a square integrable function $f(x)$ is defined as

$$T_F\{f\}(\alpha) = \frac{1}{2\pi}\int_{-\infty}^{\infty} e^{-ix\alpha} f(x)\, dx = F(\alpha)$$

Then F is square integrable and f may be recovered from F via the inversion formula

$$f(x) = \int_{-\infty}^{\infty} F(\alpha)\, e^{ix\alpha} d\alpha = T_F^{\ -1}\{F\}$$

Frequently these integrals must be evaluated by means of the residue theorem. The following is a list of Fourier transform pairs:

$I_c(x)$	$\dfrac{\sin\alpha c}{\pi\alpha}$	$c > 0$
$\dfrac{2b}{b^2 + x^2}$	$e^{-b\lvert\alpha\rvert}$	$b > 0$
$e^{-x^2/4c}\sqrt{\pi/c}$	$e^{-c\alpha^2}$	$c > 0$
$(\pi/b)\, e^{-b\lvert x\rvert}$	$\dfrac{1}{b^2 + \alpha^2}$	$b > 0$

Similarly, each piecewise continuous function $f(t)$ that is of exponential type has a Laplace transform given by

$$T_L\{f(t)\} = \int_0^{\infty} e^{-st} f(t)\, dt = F(s)$$

The function $F(s)$ is analytic for s in a half plane and f can be recovered from F via the inversion formula

$$f(t) = \frac{1}{2\pi i} \lim_{p \to \infty} \int_{c-ip}^{c+ip} e^{zt} F(z)\, dz$$

The inversion integral is generally evaluated by means of the residue theorem.

Finally, the stability of a linear system can be determined by locating the poles of the transfer function. The argument principle is often used for this purpose.

8

Potential Theory

Many problems in applied mathematics involve the determination of two dimensional vector fields, $\mathbf{V} = v_1(x, y)\,\mathbf{e}_1 + v_2(x, y)\,\mathbf{e}_2$. In a surprising number of situations these vector fields satisfy the simultaneous conditions

$$\operatorname{div} \mathbf{V} = 0 \quad \textit{and} \quad \operatorname{curl} \mathbf{V} = \mathbf{0}.$$

In such cases the theory of function of a complex variable can be applied to great advantage. Examples of two dimensional vector fields where complex variable techniques are useful include:

Static electric and magnetic fields
Gravitational fields
Steady plane irrotational flow to incompressible fluids
Steady conduction of heat

We begin this chapter by recalling a few facts from vector calculus. These results are applied to the examples listed above to show that in each case, the divergence and curl for the relevant vector field vanish simultaneously. We then show that such fields can be described in terms of a so-called complex potential function and explain how the conformal mappings methods of Chapter 6 can be used to simplify the geometry in some problems.

REVIEW OF VECTOR CALCULUS IN THE PLANE

We must begin by recalling some of the notions of differential and integral calculus of vector valued functions in the plane.

SCALAR FIELDS

The term *field* is often used as a synonym for *function* in applied mathematics. By a *scalar field* we shall mean a scalar (real) valued function $f = f(x, y)$ of the two real variables x, y. Physical quantities that are scalar fields include: density, pressure and temperature. If $f(x, y)$ is continuously differentiable with respect to both x and y in some domain D in the plane, we say f is a *smooth scalar field* in D.

VECTOR FIELDS

By the term *vector field* we shall mean a vector valued function, $\boldsymbol{V} = \boldsymbol{V}(x, y) = v_1(x, y)\boldsymbol{e}_1 + v_2(x, y)\boldsymbol{e}_2$, of the two real variables x, y. The physical quantities of velocity and force are two examples of vector fields. Here $\boldsymbol{e}_1$ and $\boldsymbol{e}_2$ denote unit vectors in the plane, parallel to the x- and y-axes, respectively. In this chapter we will deal only with plane vector fields. If the *components* $v_1(x, y)$ and $v_2(x, y)$ of the vector field $\boldsymbol{V}(x, y)$ are both continuously differentiable with respect to x and y in some domain D in the plane, then we say $\boldsymbol{V}$ is a *smooth vector field.*

VECTOR DIFFERENTIAL OPERATORS

We define a vector now whose x and y components are differential operators rather than functions

$$\nabla = \partial_x \boldsymbol{e}_1 + \partial_y \boldsymbol{e}_2 \tag{8.1}$$

We refer to ∇ as the *del-operator.* Then the del-operator can be applied to smooth scalar and vector fields in the following ways:

1. *Gradient* of a scalar field $f(x, y)$:
$$\nabla f(x, y) = \mathbf{grad} f = \partial_x f \boldsymbol{e}_1 + \partial_y f \boldsymbol{e}_2 \tag{8.2}$$
2. *Divergence* of a vector field $\boldsymbol{V}(x, y)$:
$$\nabla \cdot \boldsymbol{V} = \operatorname{div} \boldsymbol{V} = \partial_x v_1(x, y) + \partial_y v_2(x, y) \tag{8.3}$$
3. *Curl* of a vector field $\boldsymbol{V}(x, y)$:
$$\nabla \times \boldsymbol{V} = \operatorname{curl} \boldsymbol{V} = (\partial_x v_2(x, y) - \partial_y v_1(x, y))\,\boldsymbol{e}_3 \tag{8.4}$$

The vector $\boldsymbol{e}_3$ in (8.4) denotes a unit vector normal to the xy-plane. Note that the divergence of a vector field is a scalar quantity while the gradient and curl are vector fields.

PLANE CURVES

A *plane curve, C,* is a set of points of the form $\{(x, y)\colon\ x = x(t), y = y(t), a \le t \le b\}$ where $x(t), y(t)$ denote smooth functions of the single variable t. If C is a simple closed curve then $x(a) = x(b)$ and $y(a) = y(b)$ and there are no other points t_1, t_2 in the interval (a, b) where $x(t_1) = x(t_2)$ and $y(t_1) = y(t_2)$.

TANGENT AND NORMAL VECTORS

The vector, $\boldsymbol{P}(t) = x(t)\boldsymbol{e}_1 + y(t)\boldsymbol{e}_2$ will be called the position vector for the curve C. As t varies, the position vector traces out the curve C. Then it is not hard to show that the vector $\boldsymbol{P}'(t) = x'(t)\boldsymbol{e}_1 + y'(t)\boldsymbol{e}_2$, obtained by differentiating $\boldsymbol{P}(t)$, is tangent to C at each t. Similarly, the vector $\boldsymbol{Q}(t) = y'(t)\boldsymbol{e}_1 - x'(t)\boldsymbol{e}_2$, is normal to C at each t, (see Figure 8.1).

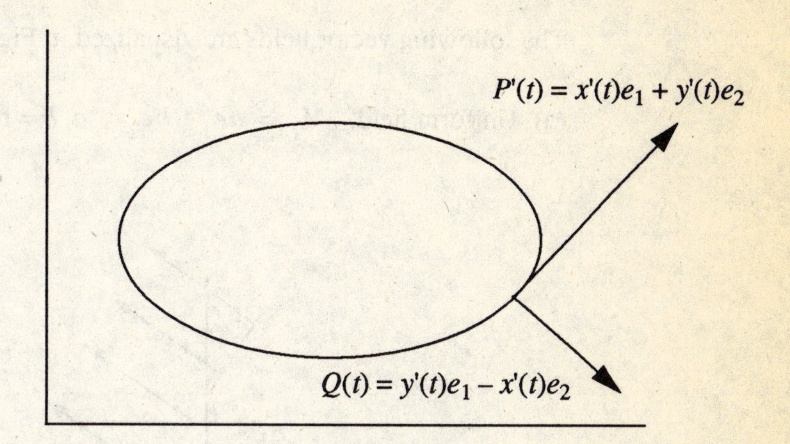

Figure 8.1

The magnitude (length) of each of the vectors $\boldsymbol{P}'(t), \boldsymbol{Q}(t)$ is equal to

$$\sqrt{x'(t)^2 + y'(t)^2} = ds/dt \tag{8.5}$$

which can be interpreted as the speed with which $P(t)$ traces out C. If we define $\boldsymbol{T}(t)$ and $N(t)$ to be unit vectors which are, respectively, tangent and normal to C at each t, then

$$\boldsymbol{P}(t) = ds/dt\boldsymbol{T}(t) \quad \text{and} \quad \boldsymbol{Q}(t) = ds/dt\boldsymbol{N}(t) \tag{8.6}$$

FLUX AND CIRCULATION

Let $V(x, y)$ be a smooth vector field in domain D in the plane and let C denote a simple closed curve in D having unit tangent and normal vectors $\boldsymbol{T}(t)$ and $N(t)$. We define the *flux of V through C* by

$$\Phi[V;C] = \int_C \boldsymbol{V} \cdot \boldsymbol{N} ds = \int_C \boldsymbol{V} \cdot \boldsymbol{Q} dt \tag{8.7}$$

The line integrals here are as defined in Chapter 3. Similarly we define the *circulation of V around C* by

$$\Gamma[V;C] = \int_C \boldsymbol{V} \cdot \boldsymbol{T} ds = \int_C \boldsymbol{V} \cdot \boldsymbol{P}' dt \tag{8.8}$$

We can interpret $\Phi[V;C]$ as the average normal component of V around C times the length of C. Similarly, $\Gamma[V;C]$ represents the length of C times the average tangential component of V around C.

EXAMPLE 8.1

Let C denote the circle
$\{(x, y): x = R\cos t, y(t) = R\sin t, 0 < t < 2\pi\}$. Then

$$P'(t) = -R\sin t e_1 + R\cos t e_2 \text{ and } Q(t) = R\cos t e_1 + R\sin t e_2.$$

The following vector fields are visualized in Figure 8.2.

(a) Uniform field: $V_1 = ae_1 + be_2 \quad a, b =$ real constants

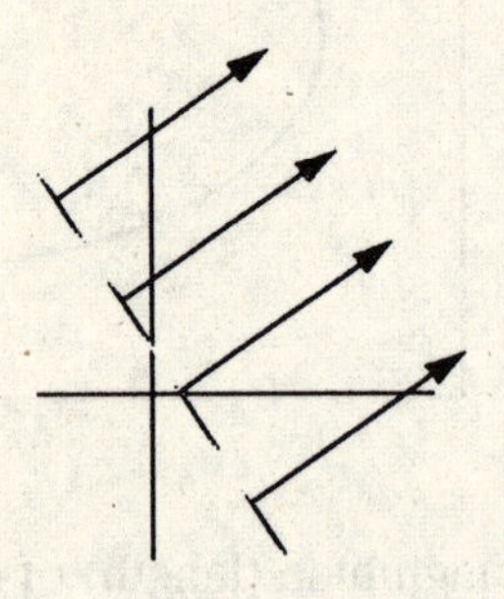

Figure 8.2(a)

(b) Source field: $V_2 = a\cos t e_1 + a\sin t e_2 \quad a > 0$

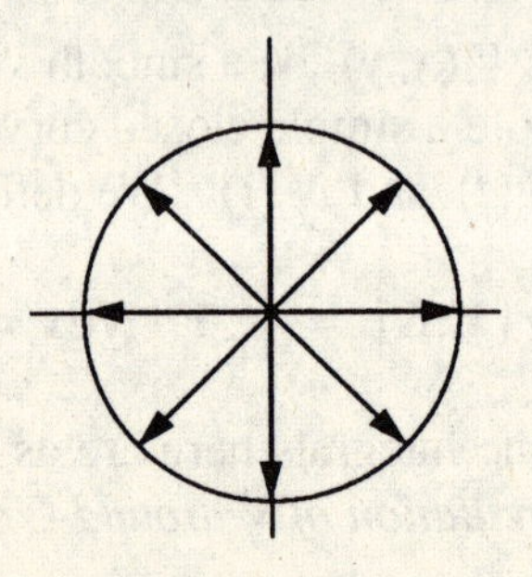

Figure 8.2(b)

(c) Vortex field: $\quad V_3 = -a\sin t e_1 + a\cos t e_2 \quad a > 0$

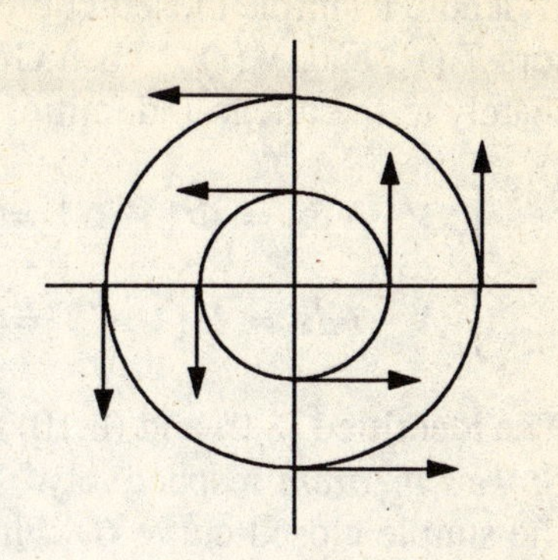

Figure 8.2(c)

Then we can easily compute:

$$\Phi[V_1;C] = \int_0^{2\pi} R(a\cos t + b\sin t)\,dt = 0,$$

$$\Gamma[V_1;C] = \int_0^{2\pi} R(-a\sin t + b\cos t)\,dt = 0;$$

i.e., there is complete cancellation of normal and tangential components around any simple closed curve C so the net flux and circulation for a uniform field is zero. For the source field we have

$$\Phi[V_2;C] = \int_0^{2\pi} aR(\cos^2 t + \sin^2 t)\,dt = 2\pi Ra,$$

$$\Gamma[V_2;C] = \int_0^{2\pi} aR(-\cos t\sin t + \sin t\cos t)\,dt = 0;$$

i.e., for a source field there is a net positive outflow through C. However, the tangential components of the source field cancel. On the other hand, for the vortex field

$$\Phi[V_3;C] = \int_0^{2\pi} aR(-\sin t\cos t + \cos t\sin t)\,dt = 0,$$

$$\Gamma[V_3;C] = \int_0^{2\pi} aR(\sin^2 t + \cos^2 t)\,dt = 2\pi Ra;$$

i.e., the normal components of the vortex field cancel out over C but the net tangential component is positive for a field that induces rotation in the counter clockwise direction.

INTEGRAL IDENTITIES

Let $V(x, y)$ be a smooth vector field in domain D in the plane and let C denote a simple closed curve in D having unit tangent and normal vectors $T(t)$ and $N(t)$. Then Green's Theorem, Theorem 3.2, leads immediately to the integral identities

$$\int_C V \cdot N ds = \Phi[V;C] = \iint_{\text{Int}\,C} \text{div}\, V dxdy \tag{8.9}$$

$$\int_C V \cdot T ds = \Gamma[V;C] = \iint_{\text{Int}\,C} \text{curl}\, V \cdot e_3 dxdy \tag{8.10}$$

The identitied (8.9) and (8.10) are versions of the *divergence theorem* and *Stoke's theorem* respectively. Here $\text{Int}\,C$ denotes the domain interior to the simple closed curve C. Finally we have the following theorems stating equivalence of certain properties of vector fields.

Theorem 8.1 Let $V(x, y)$ be a smooth vector field in the simply connected domain D in the plane. Then the following statements are equivalent:

1. $\Gamma[V;C] = 0$ for every simple closed curve C in D
2. $\text{curl}\, V = \mathbf{0}$ throughout D
3. $V = \mathbf{grad} f$ for some twice continuously differentiable scalar field f

Theorem 8.2 Let $V(x, y)$ be a smooth vector field in the simply connected domain D in the plane. Then the following statements are equivalent:

1. $\Phi[V;C] = 0$ for every simple closed curve C in D
2. $\text{div}\, V = \mathbf{0}$ throughout D
3. $V = \text{curl}\, W$ for some twice continuously differentiable vector field f

PLANE ELECTROSTATICS

An electric field E is a (vector) force field whose value at any point P is the force that would be felt by a unit charge placed at the point P. A field E that is uniform in one direction is said to be a *plane field*; i.e., $E = E(x, y)$ varies with position in the xy-plane but does not vary in the out of plane direction. Complex variable methods are useful for studying plane fields that do not vary with time. In particular, consider the following fields, defined throughout some domain D in the plane:

$\boldsymbol{E} = \boldsymbol{E}(x, y)$ the electric force field, a vector field
$Q = Q(x, y)$ the charge density field, a scalar field

FUNDAMENTAL PRINCIPLES FOR ELECTROSTATIC FIELDS

The force and charge density fields interact according to the following empirical principles:

1. For any simple closed curve C in D, $\Phi[\boldsymbol{E};C]$ is proportional to the net charge inside C; i.e., for some positive constant η,

$$\Phi[\boldsymbol{E};C] = \frac{1}{\eta}\iint_{\text{Int}\,C} Q\,dx\,dy$$

2. $\Gamma[\boldsymbol{E};C] = 0$, for any simple closed curve C in D.

Statement 1 is a version of *Gauss' law for electric fields* and 2 is a version of *Faraday's law.*

SCALAR POTENTIAL

It follows immediately from these fundamental principles, the divergence theorem, and Theorem 8.1, that

$$\eta\,\text{div}\,\boldsymbol{E} = Q \quad \text{and} \quad \boldsymbol{E} = \textbf{grad}\,\varphi \tag{8.11}$$

where $\varphi = \varphi(x, y)$ denotes a twice continuously differentiable scalar function. The function $\varphi = \varphi(x, y)$ is called the *scalar potential* associated with the vector field $\boldsymbol{E}(x, y)$. It follows from (8.11) that

$$\eta\,\text{div}\,\textbf{grad}\,\varphi = \eta(\partial_{xx}\varphi + \partial_{yy}\varphi) = Q \text{ in } D \tag{8.12}$$

COMPLEX POTENTIAL FUNCTION

Note that if $Q = 0$ throughout D, then $\varphi(x, y)$ is harmonic in D. In this case we may define a complex valued analytic function $F(z) = \varphi(x, y) + i\psi(x, y)$, called the *complex potential function* for the vector field $\boldsymbol{E}(x, y)$. Then the complex function $F(z)$ and the vector field $\boldsymbol{E}$ are related by

$$F'(z) = \partial_x\varphi(x, y) + i\partial_x\psi(x, y) = \partial_x\varphi - i\partial_y\varphi$$

$$F'(z)^* = \partial_x\varphi + i\partial_y\varphi = \boldsymbol{E}(x, y)$$

$$|F'(z)| = \sqrt{\partial_x\varphi^2 + \partial_y\varphi^2} = \|\boldsymbol{E}\|$$

EQUIPOTENTIAL CURVES AND LINES OF FORCE

The level curves of the potential function $\varphi(x, y)$ are called *equipotential curves* and level curves of the complex conjugate $\psi(x, y)$ are called *lines of force* for the field $\boldsymbol{E}$. A point where $|F'(z)| = \|\boldsymbol{E}\| = 0$ is called a *neutral point* in the field $\boldsymbol{E}$; a charge placed at a neutral point would experience no force.

EXAMPLE 8.2

(a) Consider the complex potential function

$$F(z) = -iAz = Ay - iAx \quad A = \text{real constant.}$$

Then the equipotential curves are straight lines parallel to the x-axis and the lines of force are vertical straight lines. This potential function is consistent with the electric field between two infinitely large charged flat plates that are parallel to the x-axis and extend perpendicularly from the xy-plane.

(b) For positive constant A and complex constant z_0, consider the complex potential function

$$F(z) = A\log(z - z_0) = A\ln r + iA\vartheta.$$

Then

$$\varphi(x, y) = A\ln\sqrt{(x - x_0)^2 + (y - y_0)^2},$$

$$\psi(x, y) = A\arctan((y - y_0)/(x - x_0))$$

and the equipotential curves, the level curves of φ, are circles centered at z_0. The level curves of ψ, the lines of force, are rays that originate at $z = z_0$. Thus we can identify $F(z)$ in this case as the complex potential associated with the electric field around a line charge of constant charge per unit length A. The line charge is uniformly distributed along a conductor perpendicular to the z-plane that cuts the plane at the point $z = z_0$.

PLANE MAGNETOSTATICS

A steady flow of electric current is known to induce a static magnetic field, $\boldsymbol{H}$, whose behavior is similar to that of a static electric field. If the *current density* field $\boldsymbol{J} = \boldsymbol{J}(x, y)$ is independent of time and uniform in the direction normal to the xy-plane, then the *magnetic force field*, $\boldsymbol{H} = \boldsymbol{H}(x, y)$, will be a steady, plane field. The two fields are then related by the following fundamental principles:

1. For any simple closed curve C, the average tangential component of $\boldsymbol{H}$ around C is proportional to the current flowing through C; i.e., for some positive constant μ,

$$\Gamma[\boldsymbol{H};C] = \iint_{\text{Int}\,C} \mu \boldsymbol{J} \cdot \boldsymbol{e}_3 dxdy$$

2. $\Phi[\boldsymbol{H};C] = 0$ for every simple closed curve C

Then (8.10) and Theorem 8.2 imply

$$\text{curl}\boldsymbol{H} = \mu\boldsymbol{J} \quad \text{and} \quad \boldsymbol{H} = \text{curl}\,\boldsymbol{W} \tag{8.13}$$

for some smooth vector field $\boldsymbol{W}$, called the *vector potential* associated with $\boldsymbol{H}$. The principles 1 and 2 for magnetic fields are versions of *Ampere's law* and *Gauss' law for magnetic fields*. If the current density is zero, then $\boldsymbol{H} = \textbf{grad}\,\varphi$ for some smooth scalar field $\varphi(x, y)$ and $\boldsymbol{H}$ can be associated with a complex potential function $F = F(z)$.

GRAVITATIONAL FIELDS

Consider a region of space in which mass is distributed according to a scalar *mass density field* ρ. Then the distribution of mass induces a force field $\boldsymbol{F}$ called the *gravitational force field.* If $\rho = \rho(x, y)$ is constant in time and uniform in the direction normal to the *xy*-plane then the force field is a steady, plane field. These fields are related by:

1. The average normal component of the gravitational force around any simple closed curve C is proportional to the amount of mass contained inside C; i.e., for some positive constant G

$$\Phi[\boldsymbol{F};C] = -G\iint_{\text{Int}\,C} \rho\, dxdy$$

2. $\Gamma[\boldsymbol{F};C] = 0$ for any simple closed curve C

Note that

$$\Gamma[\boldsymbol{F};C] = \int_C \boldsymbol{F} \cdot \boldsymbol{T} ds$$

may be interpreted as the *work* done by the field $\boldsymbol{F}$ in moving around the simple closed path C. Then statement 2 simply means that the field does no work in moving around any simple closed path; i.e., the field is *conservative.*

THE GRAVITATIONAL POTENTIAL

It follows from (8.9) and Theorem 8.1 applied to principles 1 and 2 that

$$\text{div}\boldsymbol{F} = -\rho G, \quad \text{and} \quad \boldsymbol{F} = \textbf{grad}\,\varphi \tag{8.14}$$

where φ denotes a smooth scalar field called the *gravitational potential.* If $\rho = 0$ then $\varphi(x, y)$ is harmonic and we may define an analytic complex potential function $f(z) = \varphi(x, y) + i\psi(x, y)$ associated with the force field $\boldsymbol{F}$. Then level curves of φ and ψ are, respectively, equipotential curves and lines of force for $\boldsymbol{F}$.

POTENTIAL FLOW OF FLUIDS

Suppose a region D in space is occupied by a moving fluid. In general we define a velocity field V, a density field ρ and a pressure field p. These fields prescribe the fluid velocity, density and pressure at each point in D and completely describe the flow. The complicated equations describing the interaction of these fields are known as the *Navier-Stokes equations.* The equations simplify considerably if we suppose:

1. ρ = constant; the fluid is *incompressible.*

2. $V = V(x, y)$; i.e., the flow is constant in time and uniform in the direction normal to the xy-plane, *steady, plane flow.*

3. $\Gamma[V;C] = 0$ for any simple closed curve in the plane; i.e., the flow is *irrotational.*

Under these assumptions, the statement of conservation of fluid mass simplifies to

4. $\Phi[V;C] = 0$ for every simple closed curve C in the plane

POTENTIAL FLOW

A flow for which assumptions 1 to 3 are satisfied is called a *potential flow.* For a potential flow, the velocity field can be determined independently from the pressure field since 3 and 4 imply

$$\operatorname{div} V = 0, \quad \text{and} \quad V = \mathbf{grad}\,\varphi \tag{8.15}$$

for some smooth scalar field $\varphi(x, y)$. Note that the *velocity potential* φ is seen to be harmonic and thus we may define an analytic *complex potential function* $F(z) = \varphi(x, y) + i\psi(x, y)$ for the velocity field V. Then

$$F'(z) = \partial_x\varphi - i\partial_y\varphi$$

and

$$\|V\| = |F'(z)| = \sqrt{\partial_x\varphi^2 + \partial_y\varphi^2}$$

A point where $\|V\| = |F'(z)| = 0$ is called a *stagnation point* of the flow.

STREAMLINES

The function ψ, a harmonic conjugate for φ, is called the *stream function* for the flow and level curves of ψ are called *streamlines.* The stream lines are the trajectories followed by particles of fluid in the flow. In fact, since the flow of fluid is along streamlines with no flow across any streamline, we can replace any streamline by a solid boundary or fluid free surface.

BERNOULLI'S LAW

For a potential flow the velocity field can be determined independently of the pressure and density fields. Nevertheless, these three fields are not unrelated. We can show that on any streamline, the velocity, density and pressure fields satisfy *Bernoulli's law*

$$\frac{1}{2}\rho\,(\partial_x\varphi^2 + \partial_y\varphi^2) + p = \text{constant along streamlines}$$

Thus the pressure in a potential flow is maximal at a stagnation point of the flow.

EXAMPLE 8.3 COMPLEX POTENTIALS

(a) Uniform Flow — For real constants V_0, α, consider the complex potential function

$$F(z) = V_0 e^{-i\alpha} z = V_0(x\cos\alpha + y\sin\alpha) + iV_0(y\cos\alpha - x\sin\alpha);$$

i.e.,

$$\varphi(x, y) = V_0(x\cos\alpha + y\sin\alpha) \text{ and } \psi(x, y) = V_0(y\cos\alpha - x\sin\alpha)$$

Then the streamlines, the level curves of ψ, are straight lines making angle α with the real axis since

$$V_0(y\cos\alpha - x\sin\alpha) = C_0 \text{ if and only if } y = x\tan\alpha + C_1$$

This is consistent with flow at constant velocity along straight lines. We refer to this as *uniform flow.*

(b) Source Flow — Consider the complex potential function

$$F(z) = A\log(z - z_0) = A\ln\sqrt{(x - x_0)^2 + (y - y_0)^2} + iA\arctan\left(\frac{y - y_0}{x - x_0}\right)$$

for A real and z_0 complex constants. Then the streamlines are rays originating at z_0; $\arg(z - z_0) = \text{Constant}$. These rays point outward or inward from z_0 according to whether A is positive or negative. We say z_0 is a *source point* if A is positive and a *sink* if A is negative. Such flow is consistent with flow from a *line source* or sink normal to the xy-plane at the point z_0. We may visualize a line source as a thin straight tube which is perpendicular to the plane and emits fluid uniformly in all directions all along its length. Note that $\Phi[V;C]$ is not zero for any simple closed curve C that contains z_0 but div V is zero at each z that is different from z_0. However, this is not in contradiction to Theorem 8.2; i.e., the velocity field V for source flow is not smooth in any simply connected domain D that contains z_0, hence the hypotheses of the theorem are not satisfied.

(c) Vortex Flow — Consider the complex potential function

$$F(z) = -iA\log(z - z_0) = A\arctan\left(\frac{y - y_0}{x - x_0}\right) - iA\ln\sqrt{(x - x_0)^2 + (y - y_0)^2}$$

for A real and z_0 complex constants. Then the streamlines of this flow are curves for which $|z - z_0|$ is constant; i.e., they are circles centered at z_0. This is consistent with a flow which is rotating around $z_0 = (x_0, y_0)$ as a center. We refer to this as *vortex flow*. Note that the circulation of the velocity field V associated with vortex flow is zero for simple closed curves C that do not enclose the vortex center, z_0 but, as we saw in Example 8.1, the circulation around a curve enclosing a vortex is not zero. Nevertheless, curl V is zero at every point z that is different from z_0. This does not violate Theorem 8.1. Rather Theorem 8.1 does not apply to vortex flow since in this case V is not smooth throughout any simply connected domain containing z_0.

STEADY STATE CONDUCTION OF HEAT

Consider a region D in space filled with a heat conducting medium. Then we may define at each point of D a scalar temperature field, T and a (vector) heat flux field, $\boldsymbol{H}$, whose value indicates the direction and rate of heat flow. We will assume that the temperature distribution is uniform in the direction normal to the xy-plane and that the temperature is constant in time. In addition, we suppose that there are no heat sources or sinks inside D. Then $T = T(x, y)$ and $\boldsymbol{H} = \boldsymbol{H}(x, y)$ satisfy:

1. *Fourier's Law of Heat Conduction*—heat flows from hot regions to cooler regions at a rate proportional to the temperature gradient; i.e. for positive constant K,

$$H(x,y) = -K\mathbf{grad}\,T(x,y).$$

2. *Steady State Condition*—the net flow of heat across any simple closed curve in the plane is zero: i.e. $\Phi[H;C] = 0$ for all simple closed curves C.

ISOTHERMAL CURVES AND LINES OF FLOW

It follows from 1 and 2 that

$$\operatorname{div} H = -K \operatorname{div}\operatorname{grad} T(x,y) = -K(\partial_{xx}T + \partial_{yy}T) = 0. \qquad (8.16)$$

Then let $S = S(x,y)$ denote the harmonic conjugate of the harmonic function T. The level curves of T are curves of constant temperature called *isotherms*. The level curves of S are a family of orthogonal trajectories to the isotherms. These are the curves along which the heat flows.

PROBLEMS IN POTENTIAL THEORY

In each of the applications just described, a complete description of the physical field is available once its complex potential function is known. In the solved problems we shall show how the complex potential for some simple geometries can be obtained by inspection. In other examples we will begin with the complex potential and show how the function can be interpreted as the potential for a situation with physical relevance. Using some simple potentials in concert with conformal mappings, we will also solve problems involving more complicated geometry.

SOLVED PROBLEMS

Plane Electrostatics

PROBLEM 8.1

Find the equipotential curves and lines of force around a cylindrical coaxial cable consisting of a cylindrical conductor at potential V_0 surrounded by a grounded cylindrical shield.

SOLUTION 8.1

We let the $a > 0$ denote the radius of the cylindrical conductor and let $b > a$ be the radius of the grounded shield (see Figure 8.3).

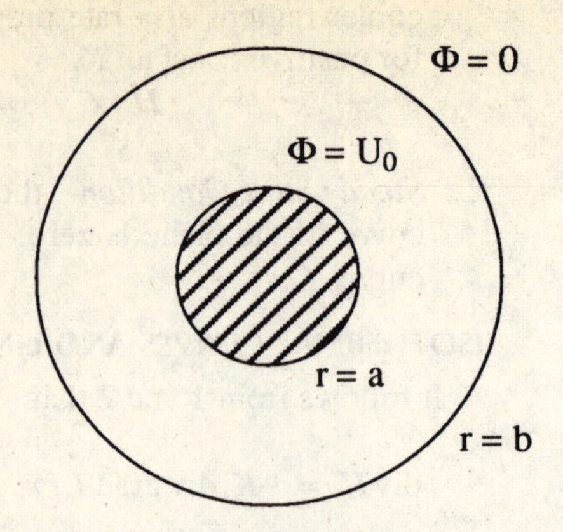

Figure 8.3

We expect a cylindrical conductor to be surrounded by circular equipotential curves which means we must find a complex analytic function $F(z)$ whose real part is a function of the radial variable only. For real constants A and B, the complex potential

$$F(z) = A\log z + B = A\ln r + B + iA\vartheta$$

has circular equipotential curves, $\varphi(r) = A\ln r + B = \text{constant}$. Then A and B must be chosen so that

$$\varphi(a) = A\ln a + B = V_0$$

$$\varphi(b) = A\ln b + B = 0;$$

i.e.,

$$A = \frac{V_0}{\ln a - \ln b} \quad \text{and} \quad B = \frac{-V_0 \ln b}{\ln a - \ln b}$$

Then

$$\varphi(r) = V_0 \frac{\ln r - \ln b}{\ln a - \ln b} \qquad a < r < b$$

is the (real) potential function in the region around the cylindrical conductor. The electric field $\boldsymbol{E}$ is obtained by computing the gradient of φ

$$\boldsymbol{E}(r) = \textbf{grad}\,\varphi = \varphi'(r)\,\partial_x r e_1 + \varphi'(r)\,\partial_y r e_2$$

$$= \varphi'(r)\,(\partial_x r e_1 + \partial_y r e_2) = \varphi'(r)\,(\frac{x}{r}e_1 + \frac{y}{r}e_2)$$

$$= V_0/(r\ln a)\,e_r$$

where

$$e_r = \frac{x}{r}e_1 + \frac{y}{r}e_2 = \text{unit vector in the radial direction.}$$

Then the direction of the electric field, and the direction of the lines of force, is the radial direction. This is also apparent from the fact that $\psi(r, \vartheta) = A\vartheta$, the conjugate of the potential function $\varphi(r)$, is constant along rays.

PROBLEM 8.2

Find the electric field around a pair of line charges, one carrying charge A and located at $z = 1$ and the other located at $z = -1$ carrying a charge of $-A$.

SOLUTION 8.2

According to Example 8.2(b), the complex potentials associated with the two line charges are $F_1(z) = A\log(z-1)$ and $F_2(z) = -A\log(z+1)$. Then the complex potential associated with the pair of line charges is equal to their sum

$$F(z) = A\log(z-1) - A\log(z+1) = A\log\frac{z-1}{z+1} \tag{1}$$

$$= A\ln(r_1/r_2) + iA(\vartheta_1 - \vartheta_2)$$

where

$$r_1 = \sqrt{(x-1)^2 + y^2} \qquad \vartheta_1 = \tan^{-1}\left(\frac{y}{x-1}\right)$$

$$r_2 = \sqrt{(x+1)^2 + y^2} \qquad \vartheta_2 = \tan^{-1}\left(\frac{y}{x+1}\right)$$

Equipotential curves $\varphi = A\ln(r_1/r_2)$ is constant along curves where $r_1/r_2 = \varphi = \text{constant}$; i.e., on curves where

$$(x-1)^2 + y^2 = \alpha^2((x+1)^2 + y^2) \tag{2}$$

We can rewrite (2) in the form

$$\left(x - \frac{1+\alpha^2}{1-\alpha^2}\right)^2 + y^2 = \frac{4\alpha^2}{(1-\alpha^2)^2} \tag{3}$$

which is the equation of a family of circles with centers on the x-axis. Note that $\varphi = A\ln 1 = 0$ on the circle corresponding to $\alpha = 1$ is the set of points $x = 0$; i.e., $\varphi = 0$ on the vertical axis. Thus we can also interpret (1) as the complex potential associated with a single line charge at $z = 1$ in the presence of a grounded vertical plane at $x = 0$.

Lines of Force $\psi = A(\vartheta_1 - \vartheta_2)$ is constant along curves where $\vartheta_1 - \vartheta_2 = \beta = \text{constant}$; i.e., where

$$\tan^{-1}(y/x+1) - \tan^{-1}(y/x-1) = \tan^{-1}(2y/(x^2-1+y^2)) = \beta$$

Here we used the identity $\tan^{-1}p - \tan^{-1}q = \tan^{-1}((p-q)/(1+pq))$. Now $2y/(x^2-1+y^2) = \tan\beta = \lambda$ if and only if

$$x^2 + (y - 1/\lambda)^2 = 1 + 1/\lambda^2 \tag{4}$$

This equation of a family of circles with centers on the y-axis. Note that the values $x = \pm 1, y = 0$ satisfy (4) for all values of λ. Thus all the circles in the family pass through both the points $(-1, 0)$ and $(1, 0)$ which means that the lines of force are circular arcs beginning at one line charge and ending at the other.

PROBLEM 8.3

Find the lines of force and equipotential curves for the electric field near the edge of a parallel plate capacitor.

SOLUTION 8.3

Consider a parallel plate capacitor with the lower plate grounded (charge = zero) and upper plate carrying charge Q. For convenience we will suppose the plates of the capacitor are separated by a distance equal to 2π.

We saw in Problem 6.17 a mapping function that carries a horizontal strip of width 2π onto the region pictured in Figure 6.15(b). We may view the region pictured in Figure 6.15(b) as the edge of the parallel plate capacitor, and we may let $w = w(z)$ denote the mapping function $z = w + e^w$ (which we know carries the strip in the w-plane onto the slit z-plane) is the inverse of this mapping function.

In Example 8.2(a) we saw that the complex potential associated with the electric field between two infinite parallel charged plates is a linear function. Thus the complex potential for infinite plates in the w-plane parallel to the u-axis has the form $F(w) = -iAw + B = Av + B - iAu$ for real constants A and B. We choose the constants A and B so that $\varphi(u, v) = Av + B$ satisfies

$$\varphi(v = -\pi) = -A\pi + B = 0 \text{ and } \varphi(v = -\pi) = A\pi + B = Q;$$

i.e.,

$$\varphi(u, v) = Q(v + \pi)/2\pi \tag{1}$$

Then $F(w) = Q(w + \pi)/2\pi$ is the complex potential function in the w-plane and if we could solve $z = w + e^w$ for w in terms of z, we could substitute this into $F(w)$ to obtain the complex potential in the z-plane. Although this is not possible we can nevertheless plot the equipotential curves and the lines of force in the z-plane as the images under the mapping

$$x = u + e^u \cos v, \quad y = v + e^u \sin v \tag{2}$$

of lines v = constant and u = constant, respectively. Figure 8.4 shows the equipotential curves. We can get some idea of the lines of force by noting that when u equals a large positive constant then e^u is much larger than either u or v and thus (2) implies that $x^2 + y^2 \approx e^{2u}$ = Constant; i.e., the lines of force for large positive u ($u >> 1$ corresponds to a location out away from the edge of the capacitor) are nearly circular. For u equal to a large negative constant, e^u is much smaller than either u or v and then (2) implies $x \approx u$, $y \approx v$; i.e., the lines of force and equipotential curves for large negative u ($u << -1$ corresponds to a location inside the capacitor, far from the edge) are vertical and horizontal lines, respectively.

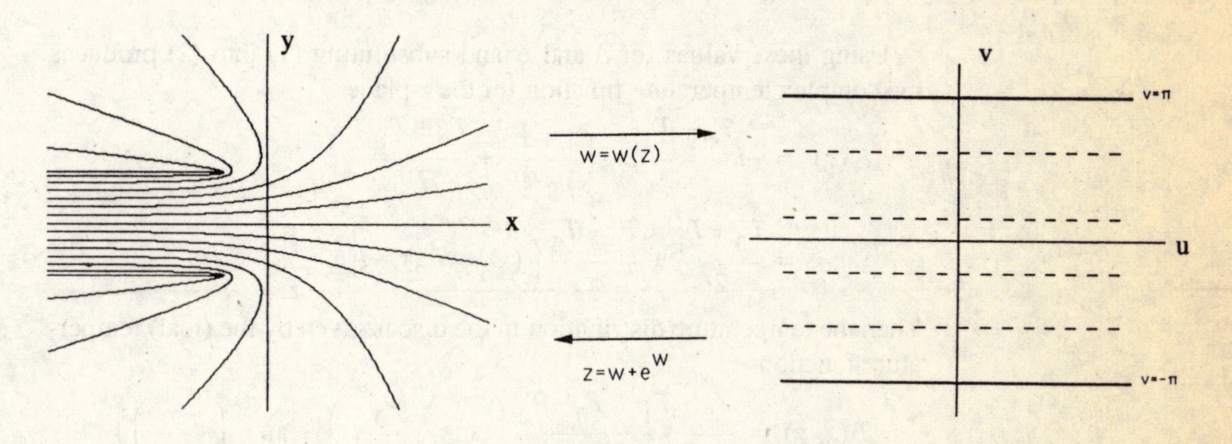

Figure 8.4

Steady State Heat Conduction

PROBLEM 8.4

A homogeneous heat conducting circular disc of radius 1 has insulated flat faces and prescribed temperatures around the circumference. Find the temperature distribution inside the disc if the upper semicircular boundary is held at the constant temperature T_0 while the lower half of the boundary is maintained at a different temperature T_1. Find the isothermal curves for this temperature distribution.

SOLUTION 8.4

The geometry of this problem can be made simpler by mapping the unit disc onto a half-plane. It is not hard to verify that the mapping

$$w = \frac{1+z}{1-z} \tag{1}$$

carries the unit disc in the z-plane onto the right half of the w-plane so that the upper semicircular half-boundary of the disc goes onto the positive v-

axis and the lower half of the boundary goes onto the negative imaginary axis in the w-plane. Then we seek a complex analytic function whose real part assumes the constant value T_1 on the negative v-axis and assumes the constant value T_0 on the positive v-axis. For real constant A and B, the function

$$F(w) = -iA\log w + B = A\arg w + B - iA\ln|w| \tag{2}$$

has these properties provided A and B are chosen such that

$$A(-\pi/2) + B = T_1 \quad \text{and} \quad A(\pi/2) + B = T_0;$$

i.e.,

$$A = (T_0 - T_1)/\pi \quad \text{and} \quad B = (T_0 + T_1)/2. \tag{3}$$

Using these values for A and B and substituting (1) into (2) produces the complex temperature function for the z-plane

$$F(z) = -i\frac{T_0 - T_1}{\pi}\log\left(\frac{1+z}{1-z}\right) + \frac{T_0 + T_1}{2}$$

$$= \frac{T_0 + T_1}{2} + \frac{T_0 - T_1}{\pi}\left((\vartheta_1 - \vartheta_2) - i\left|\frac{1+z}{1-z}\right|\right)$$

Then the temperature distribution in the disc is given by the (real) temperature function

$$T(x,y) = \frac{T_0 + T_1}{2} + \frac{T_0 - T_1}{\pi}\left(\tan^{-1}\left(\frac{y}{1+x}\right) - \tan^{-1}\left(\frac{-y}{1-x}\right)\right)$$

$$= \frac{T_0 + T_1}{2} + \frac{T_0 - T_1}{\pi}\tan^{-1}\left(\frac{2y}{x^2 - 1 + y^2}\right)$$

where we used the same simplifying identity used in Problem 8.2.

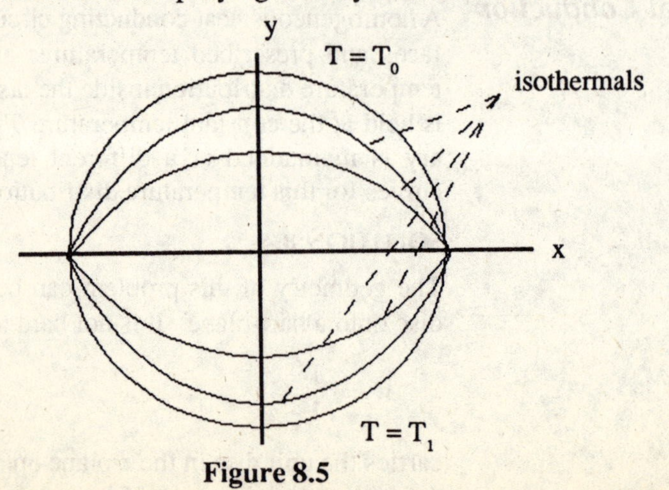

Figure 8.5

The isotherms in this problem are the same curves as the lines of force in Problem 8.2. These curves are pictured in Figure 8.5.

PROBLEM 8.5

Consider a semi-infinite homogeneous, heat conducting medium whose cross section occupies the upper half of the z-plane as shown in Figure 8.6. Suppose the portion of the boundary between $x = -1$ and $x = 1$ is insulated against the flow of heat while the portions of the boundary lying to the left and right of the insulated part are held at temperatures T_1 and T_0, respectively. Then find the isotherms and heat trajectories for this temperature distribution.

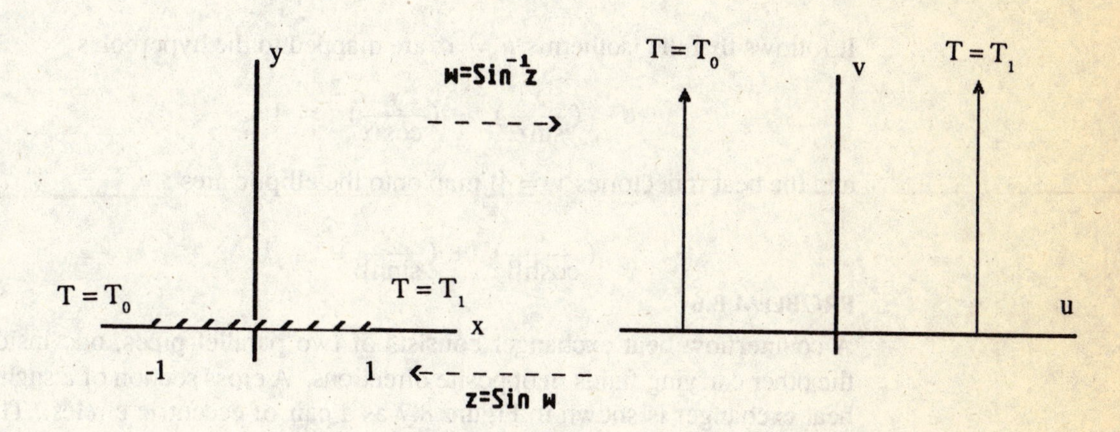

Figure 8.6

SOLUTION 8.5

In Problem 6.10 we showed that the complex sine function maps a vertical strip onto the upper half plane. Thus the inverse mapping, $w = \sin^{-1} z$, maps the upper half of the z-plane onto a vertical strip in the w-plane as shown in Figure 8.6. The segment $-1 < x < 1$ of the x-axis is mapped to the segment $-\pi/2 < u < \pi/2$ of the u-axis while the segments $x < -1$ and $x > 1$ are mapped to the vertical sides of the strip $\{u = -\pi/2, v > 0\}$ and $\{u = \pi/2, v > 0\}$, respectively.

For real constants A and B, the real part of the complex analytic function $F(w) = Aw + B$ is the harmonic function $T(u, v) = Au + B$. If we choose A and B such that

$$A(-\pi/2) + B = T_1 \quad \text{and} \quad A(\pi/2) + B = T_0$$

then T assumes the temperatures T_1 and T_2 on the vertical sides of the strip. The constants A and B are given by (3) in the previous problem. Note also that since the harmonic conjugate to $T(u)$ is the function $S(v) = Av$, the heat flows along the lines v = constant; in particular there is no flow across the bottom of the stip at $v = 0$. Thus $T = Au + B$ is the temperature distribution in the w-plane which corresponds to the prescribed conditions in the z-plane.

To find the isotherms and heat trajectories in the z-plane for this temperature distribution, we must find the images under the mapping $z = \sin w$ of the vertical lines u = constant and horizontal lines v = constant. But if

$$z = x + iy = \sin w = \sin u \cosh v + i \cos u \sinh v$$

then

$$x = \sin u \cosh v \quad \text{and} \quad y = \cos u \sinh v.$$

It follows that the isotherms $u = \alpha$ are mapped to the hyperbolas

$$\left(\frac{x}{\sin\alpha}\right)^2 - \left(\frac{y}{\cos\alpha}\right)^2 = 1$$

and the heat trajectories $v = \beta$ map onto the elliptic arcs

$$\left(\frac{x}{\cosh\beta}\right)^2 + \left(\frac{y}{\sinh\beta}\right)^2 = 1.$$

PROBLEM 8.6

A counterflow heat exchanger consists of two parallel pipes, one inside the other carrying fluids in opposite directions. A cross section of a such a heat exchanger is shown in Figure 8.7 as a pair of eccentric circles. The inner pipe, with cross sectional radius equal to $1/3$ and center at $(1/3, 0)$ in the z-plane, carries a fluid at uniform temperature T_b. The outer pipe carries a coolant and its circular cross section has radius 1 and center at the origin. Find the temperature distribution in the coolant if the boundary of the outer pipe is maintained at temperature T_a.

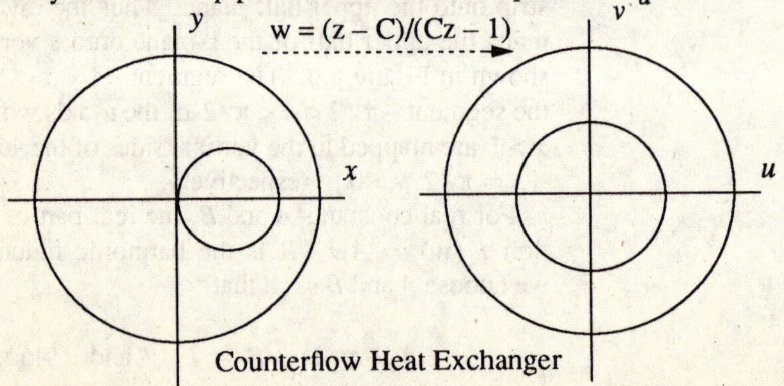

Figure 8.7

SOLUTION 8.6

In Problem 6.16 we showed that the mapping

$$w = \frac{z-C}{Cz-1} \qquad C = 7/8 \tag{1}$$

maps the heat exchanger cross section shown in Figure 8.7 onto the annular region between concentric circles of radius $1/2$ and 1 in the w-plane. Then we seek a complex analytic function whose real part depends only on the radial variable ρ, assuming the values T_b and T_a at $\rho = 1/2$ and $\rho = 1$, respectively. By analogy with the coaxial cable in Problem 8.1 we see that

$$F(w) = A\log w + B = (A\ln|w| + B) + iA\arg w \tag{2}$$

is analytic with real part $T(\rho) = A\ln\rho + B$. Then $T(\rho)$ equals T_b and T_a at $\rho = 1/2$ and $\rho = 1$, respectively, if we choose the real constants A and B as follows

$$A = (T_b - T_a)/(\ln 1/2) \qquad B = T_a. \tag{3}$$

With this choice of the constants A and B, $T(\rho) = A\ln\rho + B$ is the temperature distribution in the annular region in the w-plane. Substituting (1) into $T(\rho) = T(|w|)$, we obtain the z-plane temperature distribution

$$T(|z|) = A\ln\left|\frac{z-C}{Cz-1}\right| + B$$

$$= \frac{1}{2}A\ln\frac{(x-C)^2+y^2}{(Cx-1)^2+C^2y^2} + B \tag{4}$$

where A and B are given by (3). Clearly the isothermal curves in the z-plane are the curves where

$$\frac{(x-C)^2+y^2}{(Cx-1)^2+C^2y^2} = \alpha = \text{constant};$$

i.e.,

$$\left(x - C\frac{1-\alpha}{1-C^2\alpha}\right)^2 + y^2 = \frac{\alpha - C^2}{1-C^2\alpha} + \left(C\frac{1-\alpha}{1-C^2\alpha}\right)^2. \tag{5}$$

Equation (5) describes a family of circles with centers on the x-axis.

Potential Fluid Flow

PROBLEM 8.7

For real constants U and b, interpret the complex potential function

$$F(z) = Uz + b\log z \tag{1}$$

as the potential for:

(a) uniform flow over a blunt obstacle

(b) uniform flow over an incline

SOLUTION 8.7

In view of Example 8.3 we can see that the potential function (1) is the sum of the potentials for a uniform flow and a source flow. If we write

$$F(z) = Uz + \frac{b}{2}\ln(x^2+y^2) + i(Uy + b\tan^{-1}(y/x))$$

then we can identify the streamfunction for this potential as

$$\psi(x,y) = Uy + b\tan^{-1}(y/x) \tag{2}$$

Note that as x tends to infinity, $\tan^{-1}(y/x)$ tends to zero so the stream function $\psi(x,y)$ tends to Uy, the stream function for a uniform flow. We interpret this to mean that the flow is uniform for large x.

To visualize the flow we plot the streamlines, the curves where $\psi =$ constant. For example, the streamline where $\psi = 0$ is the set of points (x, y) such that

$$x = \frac{-y}{\tan(Uy/b)} = \frac{-y\cos(Uy/b)}{\sin(Uy/b)} \tag{3}$$

It is clear from (3) that $x(y) = x(-y)$ which implies that the streamline is a curve that is symmetric about the x-axis. We can also see from (3) that $x = 0$ when $y = \pm b\pi/(2U)$, thus $\psi = 0$ cuts the vertical axis at $y = \pm b\pi/(2U)$. Note also that x tends to infinity when y approaches $b\pi/U$ from below or approaches $-b\pi/U$ from above. This implies that $\psi = 0$ has horizontal asymptotes at $y = \pm b\pi/(2U)$.

Next we note that the Taylor series expansion for $\sin(Uy/b) = Uy/b - (Uy/b)^3/3! + \ldots$ and thus

$$x(y) = \frac{-y\cos(Uy/b)}{Uy/b - (Uy/b)^3/3! + \ldots} = \frac{-b}{U}\frac{\cos(Uy/b)}{1 - (Uy/b)^2/3! + \ldots} \tag{4}$$

We can see from (4) that $x(0) = -b/U$; i.e., the streamline $\psi = 0$ crosses the horizontal only at $x = -b/U$.

Figure 8.8 shows a sketch of the stream line $\psi = 0$. Since there is no flow across a streamline, we can interpret this curve as a solid boundary in the flow. Plotting the streamlines where ψ has a positive constant value shows that the flow has the appearance of a flow over the blunt object whose boundary is the streamline $\psi = 0$. Since $F'(z) = U + b/z$ vanishes only at $z = -b/U$, it follows that the tip of the blunt obstacle is the only stagnation point in the flow.

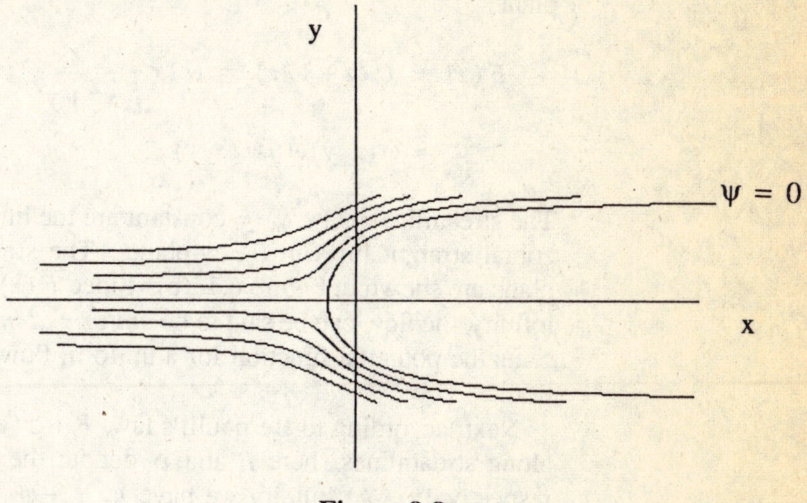

Figure 8.8

Finally, if we consider the streamline consisting of the x-axis up to the stagnation point together with the upper half of the curve $\psi = 0$, as a solid boundary for the flow, then we can interpret this flow pattern as representing uniform flow over a curved sloping beach. Note that the steepness of the slope increases as the source strength, b, is increased. When $b = 0$ there is no slope and the flow reduces to uniform flow.

PROBLEM 8.8

Find the complex potential associated to a uniform flow over a circular obstacle. Plot the streamlines for the flow and find an expression for the pressure distribution around the circumference of the circle.

SOLUTION 8.8

In Problem 6.11 we saw that the mapping function $w = z + 1/z$ maps the exterior of the unit disc in the z-plane onto the w-plane with the slit $\{-2 < u < 2, v = 0\}$ removed. Then we may view the disc in the z-plane as cylinder of circular cross section whose longitudinal axis is perpendicular to the z-plane. Similarly, the slit in the w-plane can be interpreted as a flat plate of length 4 with axis perpendicular to the w-plane. The mapping function transforms uniform flow over a cylindrical obstacle in the z-plane into uniform flow over a slit in the w-plane.

The analytic function $F(w) = Uw$ represents the complex potential for a uniform flow parallel to the u-axis in the w-plane. Since the slit is positioned along the streamline $\psi = 0$ it does not disturb the uniform flow in the w-plane. Thus $F(w)$ is the also complex potential function for uniform flow over the slit in the w-plane. Substituting $w = z + 1/z$ into the complex potential function gives the complex potential in the z-plane

$$F(z) = U(z + 1/z) = U\left(x + \frac{x}{x^2 + y^2}\right) + iU\left(y - \frac{y}{x^2 + y^2}\right)$$

$$= \varphi(x, y) + i\psi(x, y)$$

The streamlines $\psi(x, y) = \text{constant}$ are the images in the z-plane of horizontal straight lines in the w-plane. The streamlines in the upper half plane are shown in Figure 6.13(b). Since $F(z)$ tends to Uz as $|z|$ tends to infinity, the flow can be said to be *uniform at infinity* and thus $F(z)$ represents the potential function for a uniform flow around a circular obstruction.

Next, according to Bernoulli's law, $P + \rho(\varphi_x{}^2 + \varphi_y{}^2)/2$ is constant along streamlines (here P and ρ denote the fluid pressure and density respectively). At infinity we have $\varphi_x{}^2 + \varphi_y{}^2 = U^2$ and $P = P_\infty$ and therefore, by Bernoulli's law,

$$P = P_\infty + \rho(U^2 - \varphi_x{}^2 - \varphi_y{}^2)/2 \tag{1}$$

On the circumference of the circle where $z = e^{i\vartheta}$ we have

$$F'(z) = \varphi_x - i\varphi_y$$

$$= U(1 - 1/z^2) = U(1 - e^{-i2\vartheta}) = U(1 - \cos 2\vartheta) + iU \sin 2\vartheta$$

Then

$$\varphi_x{}^2 + \varphi_y{}^2 = U^2((1 - \cos 2\vartheta)^2 + \sin^2 2\vartheta)$$

$$= 2U^2(1 - \cos 2\vartheta).$$

Substituting this into (1) leads to the following expression for the pressure on the circumference of the circle as a function of the angle ϑ

$$P = P_\infty + \rho U^2(2\cos 2\vartheta - 1)/2 \tag{2}$$

Note that $F'(z)$ vanishes only at $z = \pm 1$, hence these are the only stagnation points in the flow. At $z = \pm 1$ we have $\vartheta = 0, \pi$ and thus, by (2) $P = P_{\infty} + \rho U^2/2$; i.e., the stagnation points are the points of maximum pressure. The minimum pressure occurs at $\vartheta = \pm\pi/2$ (i.e., $z = \pm i$) where we have $P = P_{\infty} - 3\rho U^2/2$. Note that the pressure given by (2) is symmetric about the x-axis so the net pressure force on the cylinder has no component in the y-direction. We interpret this to mean that uniform flow over a circular cylinder produces no *lift*.

PROBLEM 8.9

Find the pressure distribution as a function of the angle ϑ on the circumference of the unit circle for the following complex potential function

$$F(z) = U(z + 1/z) + iA\log z \tag{1}$$

Interpret this flow situation as uniform flow over a rotating cylinder.

SOLUTION 8.9

We saw in the previous problem that when $A = 0$, the potential function (1) can be interpreted as describing uniform flow at velocity U over a cylinder with axis perpendicular to the z-plane whose cross section is a unit circle centered at the origin. Adding the term $iA\log z$ to this potential amounts to superimposing a vortex flow on the uniform flow. Thus we can interpret (1) as the potential associated with a uniform flow over a rotating cylinder.

Now

$$F'(z) = U(1 - 1/z^2) + iA/z$$

implies $F'(z)$ tends to U as z tends to infinity so this flow is uniform at infinity. In addition,

$$F'(z) = 0 \quad \text{at} \quad z = -iA/2U \pm \sqrt{1 - (A/2U)^2}$$

Then when $A = 0$ the flow is uniform and the only stagnation points are at $z = \pm 1$. For $A > 0$, say $A = U$, we have stagnation points at $z = -1/2 \pm \sqrt{3/4}$ on the lower part of the unit circle. Note that these points are symmetrically located with respect to the vertical axis. For $A = 2U$, the stagnation points coalesce at $z = -i$ and for $A > 2U$ the points become separate points on the imaginary axis with one stagnation point outside the unit circle and a second, physically irrelevant point, inside the unit circle.

To compute the corresponding pressure distribution note that on the unit circle, $z = e^{i\vartheta}$, and

$$F'(z) = \varphi_x - i\varphi_y = U(1 - e^{-i2\vartheta}) + iAe^{-i\vartheta}$$
$$= U(1 - \cos 2\vartheta) + A\sin\vartheta + i(U\sin 2\vartheta + A\cos\vartheta).$$

Then

$$\varphi_x{}^2 + \varphi_y{}^2 = A^2 + 4AU\sin\vartheta + 4U^2\sin^2\vartheta$$

and

$$P = P_\infty - \frac{1}{2}\rho(A^2 - U^2 + 4AU\sin\vartheta + 4U^2\sin^2\vartheta)$$

For $A = \alpha U$, $0 < \alpha < 2$,

$$\frac{2(P - P_\infty)}{\rho U^2} = 1 - 4(\sin\vartheta + \alpha/2)^2$$

This expression for the pressure variation around the cylinder is not symmetric and generates a net upward force (lift) when integrated around the circle. This imbalance induced by the spin is what makes a curve ball curve.

PROBLEM 8.10

Find the complex potential and the stagnation points for a uniform flow over an inclined flat plate.

SOLUTION 8.10

We shall suppose the flat plate occupies the interval $-2 < x < 2$ on the x-axis and the flow is inclined at an angle α to the x-axis. Now the results of Problem 8.8 indicate that the potential function for a uniform flow over a circular cylinder in the w-plane is $F(w) = U(w + 1/w)$. Then it is not hard to show that the potential function for an inclined uniform flow is

$$F(w) = U(e^{-i\alpha}w + e^{i\alpha}/w)$$

where α denotes the angle of inclination. Now the mapping $z = w + 1/w$ carries the unit circle in the w-plane to the z-plane with the slit $\{-2 < x < 2\}$ removed. Then since $z = w + 1/w$ implies that

$$w = (z + \sqrt{z^2 - 4})/2$$

we conclude that the potential function of the flow in the z-plane is

$$F(z) = U\left(e^{-i\alpha}(z + \sqrt{z^2 - 4})/2 + \frac{2e^{i\alpha}}{z + \sqrt{z^2 - 4}}\right)$$

$$= U\,(z\,(e^{i\alpha}+e^{-i\alpha})/2-\sqrt{z^2-4}\,(e^{i\alpha}+e^{-i\alpha})/2)$$

$$= U\,(z\cos\alpha - i\sqrt{z^2-4}\sin\alpha)\,.$$

Then

$$F'(z) = U\left(\cos\alpha - i\sin\alpha\frac{z}{\sqrt{z^2-4}}\right)$$

and $F'(z) = 0$ if

$$\frac{z^2-4}{z^2} = -\tan^2\alpha;$$

i.e.,

$$z = \pm 2\cos\alpha.$$

Thus the stagnation points, the points of maximum pressure, are located at $z = \pm 2\cos\alpha$. These points are on opposite sides of the plate and thus the flow exerts a moment on the plate.

SUMMARY

The vector field E is called a potential field if $\mathbf{E} = \mathbf{grad}\,\varphi$ *for some smooth function* φ. *If E has the additional property that* $\operatorname{div}\mathbf{E} = 0$, *then* φ *is seen to be harmonic. Hence if* $\mathbf{E} = \mathbf{E}(x, y)$ *is a two dimensional field then* $\varphi = \varphi(x, y)$ *is the real or imaginary part of an analytic function F of the complex variable* $z = x + iy$. *The connection between the analytic function* $F = \varphi + i\psi$ *and the vector field* $\mathbf{E} = \mathbf{grad}\,\varphi$ *is explained by:*

(a) For $F(z) = \varphi(x, y) + i\psi(x, y)$ *analytic,* φ *and* ψ *are harmonic conjugates, hence the level curves of* φ *are everywhere orthogonal to the level curves of* ψ

(b) The level curves of any smooth function φ *are everywhere orthogonal to the vector field* $\mathbf{grad}\,\varphi$

Thus the level curves of ψ *are everywhere tangent to the vector field* $\mathbf{grad}\,\varphi = \mathbf{E}$. *In addition* $F'(z) = \partial_x\varphi - i\partial_y\varphi$ *so the modulus of* F' *equals the magnitude of E.*

Static electric and magnetic fields, steady state temperature fields and the velocity field for steady irrotational flow of an incompressible fluid are all examples of two dimensional vector fields for which the curl and divergence are both zero. Hence each has a complex potential

Electric Field	$F(z) = \varphi(x, y) + i\psi(x, y)$ *complex electric potential*
	$\mathbf{E} = \mathbf{grad}\,\varphi$ *electric force field*
	curves of constant φ: *equipotential curves*
	curves of constant ψ: *lines of electric force*
Magnetic Field	$F(z) = \varphi(x, y) + i\psi(x, y)$ *complex magnetic potential*
	$\mathbf{H} = \mathbf{grad}\,\varphi$ *magnetic force field*
	curves of constant φ: *equipotential curves*
	curves of constant ψ: *lines of magnetic force*
Steady Heat Conduction	
	$F(z) = T(x, y) + iS(x, y)$
	$\mathbf{H} = -K\,\mathbf{grad}\,T$ *heat flux field*
	curves of constant T: *isothermal curves*
	curves of constant S: *heat flow trajectories*
Potential Fluid Flow	
	$F(z) = \varphi(x, y) + i\psi(x, y)$ *complex velocity potential*
	$\mathbf{V} = \mathbf{grad}\,\varphi$ *fluid velocity field*
	curves of constant ψ: *streamlines of the flow*
	(note: any streamline can be interpreted as a solid boundary for the flow)

Problems in potential theory may be approached directly or indirectly. In the indirect approach we choose a complex analytic function $F(z)$ *and sketch the level curves for the real and imaginary parts to see whether this potential is descriptive of any situation with physical meaning. This indirect approach is illustrated in Problem 8.7.*

Alternatively, we may start with a physical problem for which the complex potential function is initially unknown and seek to determine $F(z)$ *directly. For example, we may use conformal mapping methods to transform the geometry of the situation for which the complex potential* $F(z)$ *is unknown into a situation in which the complex potential is already known. Then the original potential is found by inverting the mapping. Problem 8.10 is a problem of this type.*

Index

D

E

F

G

H

P

Q

R

S

T

U

V

Z